Python for Agriculturists

Muhammad Azhar Iqbal

Python for Agriculturists

Springer

Muhammad Azhar Iqbal
University of Leeds
Leeds, UK

ISBN 978-3-032-01441-2 ISBN 978-3-032-01442-9 (eBook)
https://doi.org/10.1007/978-3-032-01442-9

This Springer imprint is published by the registered company Springer Nature Switzerland AG
The registered company address is: Gewerbestrasse 11, 6330 Cham, Switzerland

With profound gratitude to my Father, Mother, Brothers, and Sisters whose steadfast support made my education possible and my Wife whose support has made this work possible.

Preface

In recent times, agriculture is not just about working with soil, seeds, water, fertilizer, pesticide, etc. within agricultural fields, but it is also about using data and information and communication technologies (ICT) to help agriculturists in making better decisions on today's key challenges, i.e., how to produce more food, tackle climate change, conserve farm resources, and manage farms in ways that protect the environment. This is where digital agriculture plays a crucial role, offering solutions through the use of computing devices and computer programs.

This book, *Python for Agriculturists*, is designed for university students in agriculture programs to help them learn the use of the Python Programming Language, which is a popular, beginner-friendly, and easy-to-learn programming language, particularly for applying data analysis in agricultural work. Recognizing the limited understanding of agricultural students (especially in developing countries) about computer programming, I realized the need to introduce them to the basic principles and key concepts of (Python) programming to solve agricultural tasks/problems and prepare them to play their role in modern digital agriculture. To the best of my knowledge, this is the first programming book written specifically with the problems and context of core agricultural branches (i.e., agronomy, entomology, plant breeding and genetics, plant pathology, forestry, food technology, animal husbandry, veterinary sciences, etc.) in mind. A key feature of this book is that it ultimately guides agriculturists to understand how Python can be applied across all phases of agricultural data processing, i.e., from storing and reading data to preprocessing, analyzing, and extracting useful insights. You do not need to be a computer expert to benefit from this book; it takes you step by step, introducing you to Python's basics to more advanced constructs and showing how Python can help manage, process, and analyze agricultural data.

In Chap. 1, you will learn what digital agriculture is, why it matters, and how knowing computer programming is becoming important in modern farming. This chapter also explains the basics of computer programming, introduces Python, and shows the use of various Integrated Development Environments (i.e., IDLE, PyCharm, and VS Code) that can help you write and execute Python programs.

Chapter 2 explains how to write simple Python programs through the use of elementary Python programming concepts and components, i.e., the use of comments, data types, indentation, performing calculations, taking user input, showing results output, and handling errors.

In Chap. 3, you will learn how to write your programs using decision-making (conditional) statements and (iterative) loops, to respond to different situations and repeat tasks, respectively.

Chapter 4 is about functions/methods and modules, which help you to break or divide your large-scale program into smaller, reusable pieces that are easier to manage.

Chapter 5 covers Python's core data structures (i.e., lists, tuples, sets, and dictionaries) for storing collections of information and how this can be processed to get useful insight.

In Chap. 6, you will explore how to read from and write to files, so you can work with agricultural data (i.e., weather records or crop reports) available on personal and public computing devices.

Finally, Chaps. 7 and 8 introduce some of the most useful Python Packages (i.e., NumPy for working with numbers and Pandas for working with tables of data) and Python Libraries (i.e., Matplotlib for making charts/graphs and Scikit-learn for doing machine learning). Ultimately, these packages and libraries will help you analyze agricultural data, find patterns, and make better decisions.

Leeds, UK Muhammad Azhar Iqbal

Competing Interests The author has no competing interests to declare that are relevant to the content of this manuscript.

Contents

1 **Preliminaries** 1

2 **Basic Structure and Elementary Components of Python Programs** 27

3 **Control Flow Structures/Statements** 61

4 **Python Functions, Methods, and Modules** 95

5 **Python Data Structures** 137

6 **File Handling in Python** 183

7 **Data Science and Python Packages** 205

8 **Data Science and Python Libraries** 255

Index 303

About the Author

Muhammad Azhar Iqbal completed his PhD in Communication and Information Systems in 2012 from Huazhong University of Science and Technology (HUST), China. Later, he held various positions at different universities. Prior to his current tenure as Assistant Professor at the University of Leeds, Dr. Azhar served as a Teaching Fellow at Lancaster University (LU, United Kingdom), an Associate Professor at Southwest Jiaotong University (SWJTU, China), and an Associate Professor at Capital University of Science and Technology (CUST, Pakistan). He has received senior membership with IEEE and a fellowship of the Higher Education Academy. He has authored several international conferences and journal publications and is the lead author of three books on computer network simulations, Internet of things (IoT), and digital agriculture. Dr. Azhar's current research focuses on the field of agricultural digitalization, with a strong focus on developing artificial intelligence-driven solutions that improve the sustainability of crop and animal production systems.

Chapter 1
Preliminaries

Before starting with Python programming, agriculturists must grasp some fundamental concepts of Digital Agriculture and Computer Programming. An overview of Digital Agriculture, along with an understanding of basic Computer Programming principles, will lay the groundwork for effectively utilizing Python in the modern agricultural context.

1.1 Overview of Digital Agriculture

Agriculture serves the fundamental demands and necessities of humans with the provisioning of food (i.e., crops, vegetables, fruits, herbs, spices, beverages, plants, meat, milk, eggs, honey, oil, etc.), clothing (i.e., cotton, wool, silk, leather, etc.), and shelter (i.e., lumber, carpeting, plastics, etc.). Likewise, it helps humans with medicine, household items, fuel, and recreation [1]. The word Agriculture is derived from the Latin word *Agricultura*, which is ultimately a combination of two Latin words *Ager* (meaning land or field) and *Cultura* (meaning cultivation or growing). Thus, in the literal sense, the word Agriculture means the cultivation of land. However, agricultural study is concerned not only with crop cultivation and planting trees but also with rearing livestock [2, 3]. In the twenty-first century, agriculture faces three primary challenges: the impacts of extreme weather and climate variability, the rapid growth of the global population, and the reduction of arable land due to urbanization [4]. In addition to these primary challenges, other key factors driving the adoption of digital technologies in agriculture include the need for remote farm monitoring and management for higher productivity at low cost and reduced environmental impact [5]. Therefore, in today's world, the agriculture sector and its associated fields are increasingly shifting toward adopting digital farming practices, particularly through the use of advanced Information and Communication Technologies (ICT).

M. A. Iqbal, *Python for Agriculturists*,
https://doi.org/10.1007/978-3-032-01442-9_1

1.1.1 Definition and Scope of Digital Agriculture

The German Agricultural Society (Deutsche Landwirtschafts-Gesellschaft (DLG)) [6] has defined digital farming as "*consistent application of the methods of precision farming and smart farming, internal and external networking of the farm and use of web-based data platforms together with BigData analyses*" [7]. According to this definition, digital farming

- is the combination of two advanced agricultural farm management concepts, i.e., precision farming and smart farming
- involves the connectivity of ICT devices (i.e., sensors, drones, smartphones, etc.) with internal farm entities to collect data across the agricultural farm
- involves the sharing of farm data (stored on cloud-based platforms through web-based systems) to connect external farm entities, i.e., suppliers, markets, research institutions, government bodies, etc.
- includes the analysis of massive available agriculture and environmental data for different purposes (i.e., weather predictions, market trends, pest forecasts, etc.) to facilitate agriculture stakeholders

Therefore, the scope of Digital Agriculture is vast and empowered by advanced digital technologies and computing techniques (i.e., sensors, drones, data storage, data analytics using artificial intelligence (AI) or machine/deep learning approaches) to support various aspects of agriculture, i.e., monitoring and management of crops/livestock, optimization of farm resources, automation of farm machinery and irrigation systems, promotion of supply chain transparency and sustainability while minimizing environmental impacts. This trend of agricultural digitalization brings innovations to all branches of agriculture (mainly classified into five categories, i.e., Crop/Plant/Tree Production and Management, Crop/Plant/Tree Improvement, Crop/Plant/Tree Protection, Animal/Livestock Farming, and Allied Disciplines [3] as shown in Fig. 1.1) and helps agriculturists to work more precisely, efficiently, and sustainably. In addition, it improves decision-making and paves the way for a future characterized by a more data-driven and adaptable approach to farming within the Digital Agriculture Ecosystem.

1.1.2 Digital Agriculture Ecosystem

A Digital Agricultural Ecosystem is an adaptive, scalable, and sustainable framework composed of a network of interconnected digital resources that create an informative environment to support the automation of agricultural practices while promoting collaboration among agricultural stakeholders, even in the absence of established partnerships or strong connections [8]. The primary goal of the digital agricultural ecosystem is to enhance the quantity and quality of agricultural production by sustainably integrating digital technologies. In addition, reducing labor costs

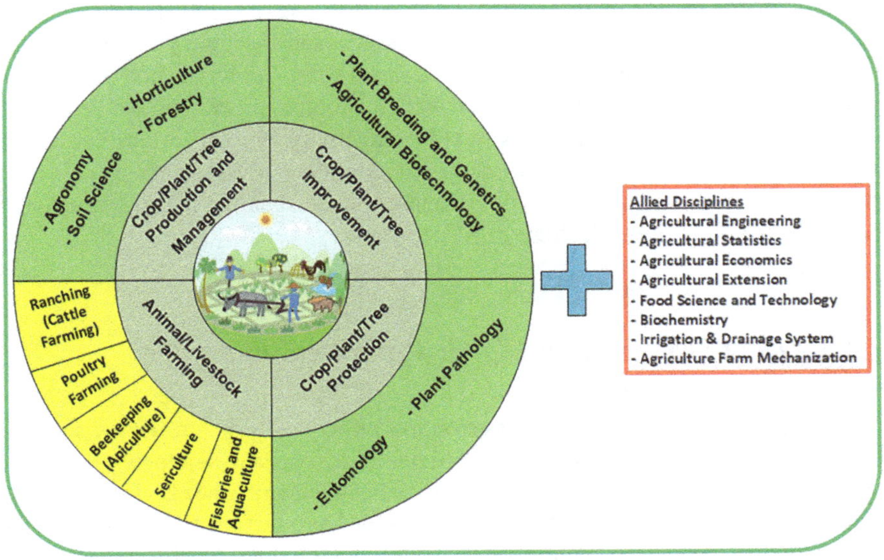

Fig. 1.1 Classification of the branches of agriculture

and boosting income, these ecosystems enable farmers and farming communities to make informed decisions that also improve both the quantity and quality of their agricultural products. The four key components of this ecosystem include agricultural stakeholders, farming machinery/equipment, agricultural operations, and digital technologies [1]. *Agricultural Stakeholders* include farmers, ranchers, seed providers, pesticide and fertilizer companies, agricultural consultants, retailers, advisors, agricultural extension workers, aggregators, distributors, marketers, transporters, animal feed companies, veterinarians, agricultural credit, and financial institutions, etc. *Agricultural Machinery* includes combine harvesters, forage harvesters, tractors, and tractor attachments (plows, harrows, seeders, fertilizer spreaders, sprayers, mowers, cultivators, balers, wagons or trailers, orchard cabs, etc.). *Agricultural operations* at different phases of crop/animal production include *Pre-cultivation* operations (i.e., land preparation, planting, sowing, etc.), *Cultivation* operations (i.e., irrigation, fertilization, spraying pesticide, weeding, plant counting, yield prediction, monitoring and management of crops, crop fields, and farm resources, etc.), *Harvesting* that includes crop harvesting along with labor and machinery management, and *Postharvest handling* (i.e., transportation and storage of agricultural products, food packaging, food processing, food safety or preservation, food distribution, food quality management, and food marketing, etc.). *Digital Technologies* include all electronic devices and computing techniques that are mainly part of the IoT (Internet of Things) technology stack including sensors, drones, wired and wireless communication, data/cloud storage, and data analysis using various AI techniques, i.e., machine learning and deep learning.

The implementation of IoT technologies enables ongoing real-time monitoring and automated management of farming operations and machinery on agricultural land. It also helps reduce the workload for farmers while optimizing the use of farm resources through cost-effective data collection, storage, processing, and analysis, ultimately enhancing the farm's expected productivity [9]. Since data forms the backbone of any digital ecosystem, understanding the various stages of data processing is crucial for gaining a comprehensive understanding of how functional components within the agriculture ecosystem work.

1.1.3 Data Processing in Digital Agriculture Ecosystem

Digital agricultural ecosystems operate on IoT principles, which rely on a network of interconnected digital resources, smart devices, advanced technologies, and computational methods. These elements work together to enhance the efficiency of agricultural processes, ultimately leading to improved production and higher quality agricultural outputs. The digitalization of farming activities within an IoT-driven agricultural ecosystem progresses through several stages of data processing, including Data Acquisition, Data Transmission, Data Storage, and Data Analytics [1], as shown in Fig. 1.2. The specifics of these data processing stages related to agriculture are outlined below.

1.1.3.1 Data Acquisition

The data acquisition phase involves gathering a wide range of agricultural data critical for improving crop and livestock management. For crops and plants, this includes data on soil moisture, soil temperature, soil texture, climate conditions, temperature, humidity, pest infestations, crop diseases, fertilization and irrigation needs, and plant growth patterns. In terms of livestock production, the data collected focuses on animal movements, grazing patterns, rumination, and overall health and well-being, such as monitoring heart rate, heat, and stress levels. This data is obtained using various digital technologies such as RFID (Radio-Frequency Identification) tags, RGB and multispectral cameras, Unmanned Aerial Vehicles (UAVs), satellites, GPS (Global Positioning System), mobile phones, and a wide array of sensors, including pH sensors, gas sensors, ultraviolet sensors, rainfall sensors, motion detectors, soil moisture sensors, temperature sensors, humidity sensors, and barometric pressure sensors. These technologies enable precise data collection, which is crucial for optimizing agricultural operations.

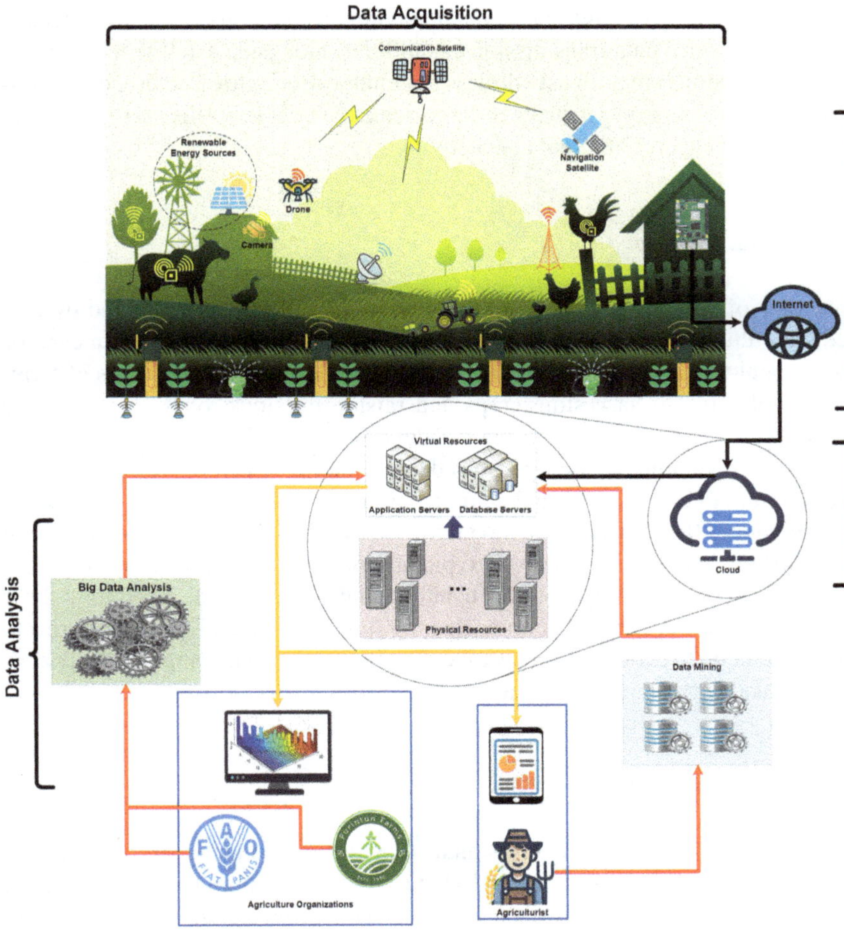

Fig. 1.2 IoT-based digital agricultural ecosystem

1.1.3.2 Data Transmission

The transmission of agricultural data (including vital field parameters such as soil, crops, livestock, and weather) from field sensors and devices is primarily reliant on wireless communication infrastructure. Depending on the connectivity range, various advanced wireless technologies used within Digital Agriculture Ecosystems can be classified into several categories. These include *short-range communication technologies* such as RFID, NFC (Near Field Communication), Bluetooth, and ZigBee, as well as *long-range communication technologies* such as LoRaWAN (Long Range Wide Area Network), NB-IoT (Narrowband IoT), and SigFox. Additionally, improved *local area communication* is supported by technologies such as WiFi 6, while *cellular communication* is facilitated by networks such as 2G/3G and LTE

(Long-Term Evolution). *High-band communication technologies* (including 4G and 5G) further enhance data transmission capabilities, alongside satellite telecommunications. In addition to a robust wireless communication infrastructure, data transfer from edge routers or gateways to large-scale data centers relies on *broadband technology* that ensures efficient connectivity.

1.1.3.3 Data Storage

The storage of vast amounts of data (often referred as BigData) generated by IoT-based agricultural applications is managed through large-scale cloud data centers. These agricultural applications provide services that encompass all stages of farming, from planting to harvesting crops and raising livestock, on both small- and large-scale farms. Various sensors and IoT devices deployed in these processes produce a massive volume of heterogeneous data, including structured (clearly defined in the form of tables with rows and columns), semi-structured (partially organized), and unstructured (lacks predefined structure) formats. Storing such a vast amount of data on locally available hardware is typically impractical, and conventional data management tools are inadequate for handling and processing it. To address these challenges, *cloud storage* and *cloud computing* offer extensive IT infrastructure and ultimately enable IoT systems to meet the flexible storage and computational needs of farmers and agricultural organizations [10].

1.1.3.4 Data Analytics

Data analytics in agriculture involves analyzing stored data using various computing technologies, including image and video processing, computer vision, pattern recognition, machine learning, deep learning, audio conversion, and natural language processing. Accurate data analysis in farming is essential for enhancing productivity by improving operational efficiency. Different analytical techniques help to boost both the quantity and quality of agricultural products by optimizing resource usage through identifying key environmental factors, such as humidity and temperature, which ultimately reduce farm management costs [11, 12]. These techniques also aid agriculturists in predicting extreme conditions such as floods, droughts, and diseases, allowing farmers to take preventive actions swiftly. Additionally, early detection of pests and crop diseases during crop development enables timely protective measures. Agricultural departments, organizations, and companies at both private and government levels can also utilize this data to make informed decisions, assisting in short, medium, and long-term planning, policy development, and marketing strategies.

All these stages directly or indirectly expect the awareness of computing knowledge from modern-day agriculturists.

1.2 Significance of Computing Knowledge in Digital Agriculture

The computing domain (encompassing both the knowledge of computer hardware and computer software (programs)) involves understanding how computer systems function. In today's digital age, the knowledge of computers and programming is increasingly necessary for modern agriculturists due to the growing reliance on technology to optimize and innovate farming practices. Hence, this knowledge will ultimately enable them to use technology in a better way to improve efficiency, productivity, and sustainability while adapting to changing environmental and market conditions [13]. Below are key reasons why knowledge of computing is essential for agriculturists.

1.2.1 Farm Management Software

Modern farm management platforms integrate numerous data streams to help farmers to analyze farming activities. Agriculturists with basic computer and programming skills can configure, customize, or even create their own farm management software to suit their specific needs, giving them more control over inventory, labor, logistics, and finances.

1.2.2 Automation of Farm Machinery

Many agricultural tasks (i.e., planting, monitoring, and harvesting) are now performed by automated machines and robots, which are programmed to perform specific tasks. Understanding the basics of computing and programming languages can help agriculturists to configure these systems, troubleshoot issues, and develop solutions to their specific needs, leading to increased efficiency and productivity.

1.2.3 Data Collection and Analysis

Modern farming involves collecting vast amounts of data from various sources such as soil sensors, weather stations, satellite imagery, drones, and so on. A better understanding of how to use computers and programming allows agriculturists to analyze this data to identify trends and make informed decisions to improve crop yields, resource management, and mitigate risks.

1.2.4 Artificial Intelligence (AI) and Machine/Deep Learning (ML/DL) Applications

The knowledge of computers and programming enables agriculturists to implement and customize different AI, ML/DL models (used to predict weather patterns, disease outbreaks, and even market trends) and ultimately help them to understand, predict, and optimize outcomes in areas like crop yield, pest control, and water management.

1.2.5 Supply Chain and Blockchain Technology

In recent times, blockchain is being used to ensure traceability and transparency in the agricultural supply chain, from seed to consumer. Understanding programming and blockchain technology allows agriculturists to integrate these systems in organic and specialty markets to improve trust and accountability.

1.2.6 Mobile and Web-Based Applications

Many agriculturists rely on mobile apps for real-time data on weather, market prices, and crop health. Therefore, fundamental computing (and programming) knowledge enables agriculturists to understand how these apps work and how custom applications can be developed to better meet their needs.

1.2.7 Integration of Internet of Things (IoT) Devices

IoT devices are transforming agriculture by connecting sensors, drones, and other equipment that monitor the environmental conditions of agricultural fields in real-time. Agriculturists who understand computing and programming can customize these systems, ensuring the right data is being collected and interpreted, leading to better resource management and timely responses to crop or livestock needs.

1.2.8 Smart and Precision Agriculture

Smart and Precision farming relies on technologies that enable farmers to monitor and control farming operations from planting to harvesting. Agriculturists with knowledge of computing and programming can better utilize tools that control automated machinery, manage variable rate applications, and optimize irrigation, fertilization, and pesticide use, leading to cost reduction and environmental sustainability.

1.2.9 Sustainability and Environmental Impact

Computer simulations and models help agriculturists to forecast the environmental impact of their farming practices. Programming skills allow agriculturists to customize models for soil conservation, water management, and carbon footprint reduction and thus supporting more sustainable farming practices.

1.2.10 Global Competitiveness

As agriculture becomes more advanced (technologically) and therefore staying competitive requires an understanding of modern computing tools/systems and programming languages. Knowledge of computing and computer programming helps agriculturists to keep up with industry innovations and increase their ability to compete in a global market by adopting new practices to earn more profit.

These key reasons highlight that to effectively apply computing knowledge in the field of agriculture, it is essential for modern agriculturists to understand computer programming that is precisely the purpose for which this book has been written. The core theme of this book revolves around learning how to write computer programs in Python to solve problems related to agriculture, for which an in-depth knowledge of computer hardware is not required. Therefore, skipping the details of computer hardware components, a direct discussion of software, computer programs, and programming languages has been given below.

1.3 Fundamentals of Computer Programming

A computer is an electronic device consisting of several hardware components (i.e., CPU, memory, storage, communication devices, motherboard, etc.) that can accept/store input in the form of digitalized data and process it to generate specific output based on an intangible set of instructions known as a computer program (that is one of the components of software).

1.3.1 Software and Computer Program

The terms Software and Computer Program are sometimes used interchangeably, but they are not exactly the same and the difference lies in their scope and purpose. For example, Software is a broader term that encompasses not only the computer program but also all the associated libraries, files, resources, and documentation to support its operations. Therefore, Software consists of multiple programs that work together and include everything necessary to execute and manage tasks on a computer. Examples of Software include operating systems (e.g., Windows, UNIX, Linux, etc.), office suites (including MS Word, Excel, Access, Visio, etc.), mobile applications, and so on. On the other hand, a computer program is a standalone or self-contained specific set or sequence of instructions that accomplishes a specific task or produces specific results. For example, a set of instructions calculating the sum of multiple numbers can be considered as a standalone computer program. Computer programs are written in a computer programming language.

1.3.2 Computer Programming Languages

Computers are unable to comprehend human languages; therefore, computer programs are required to be written in a language that computers can understand or interpret. A computer's native language is Machine Language that consists of a set of built-in primitive instructions in the form of binary code (0s and 1s) and is directly understood by a computer's hardware. However, Machine Language is not preferred today because computer programs written in it are difficult to write, read, and debug. Moreover, Machine Language is hardware-specific and therefore a deep understanding of hardware is required to write programs in it. These days, developers or programmers prefer high-level languages to write programs and develop software because these languages are user-friendly, close to human languages, and easier to write, read, and debug. Moreover, these languages are portable (not hardware-specific) and allow programmers to write code more efficiently, with fewer errors, and make the development process much faster. Here, it is important to understand the concepts of a Compiler and an Interpreter. As high-level languages are not understandable by computing devices, so a special software, generally known as a translator, is required to translate the high-level language to machine-level code. Two types of translators are a *Compiler* and an *Interpreter*. The differences between a *Compiler* and an *Interpreter* have been shown in Table 1.1. Moreover, a brief list of popular high-level computer programming languages (along with their translator-based classification as compiled or interpreted), is shown in Table 1.2

Among different high-level programming languages, these days Python is mostly preferred over other programming languages because of its simplicity, ease of learning, versatility to develop various applications, and vast library support for machine learning applications.

Table 1.1 Difference between *Compiler* and *Interpreter*

Feature	Compiler	Interpreter
Definition	Software tool that translates an entire computer program before actual execution	Software tool that translates and executes the program line by line
Memory	Requires more memory to store machine code information in a special file (known as an object file)	Requires less memory as machine code is not required to be stored in the object file
Speed	In general, faster translation of the whole code	Slower than the compiler due to line-by-line execution
Error detection	Detect and report errors after compiling the whole code	Detect and report errors line by line during execution
Examples	C and C++	JavaScript and Python

Table 1.2 Popular high-level computer programming languages

Language	Description	Translator
C++	A powerful language used to develop system and application software	Compiler-based
Java	A cross-platform language used to develop web and mobile applications	Both Compiler and Interpreter-based
JavaScript	A robust language used for client-side and server-side web development	Interpreter-based
C#	A Microsoft-developed language used to develop both desktop applications and web applications	Compiler-based
Python	A powerful language that is widely used in data science and AI applications	Interpreter-based

1.4 Python Programming Language: History and Importance

Python programming language is created by Guido van Rossum in 1990. He started it as a hobby project, but now it has grown into a widely used programming language in both industry and academia. Python is a simple, open-source, interpreted, versatile, platform-independent, and general-purpose language. Its flexibility in supporting multiple programming paradigms allows developers to write Python code for a wide range of tasks across different programming approaches. These characteristics of the Python language also make it ideal for beginners to learn, while its efficiency supports the development of complex applications. Moreover, Python consistently ranks as a top choice for programmers doing work in the machine learning domain [14, 15].

To ensure its continuous improvement, the Python Software Foundation (PSF) is currently responsible for the development and maintenance of the Python programming language. Since Python is open-source, it allows developers from around the world to contribute to its development, resulting in a highly active and collaborative community that provides continuous support, resources, and improvements. Presently, two main versions of Python coexist, i.e., Python 2 and Python 3, but they are not compatible, though there are tools to convert Python 2 code to Python 3 syntax. In this book, all example programs and code snippets have been written using Python 3.

1.4.1 Features of Python Programming Language

Details about the features of the Python programming language are discussed below [15, 16].

1.4.1.1 Free and Open-Source

Python is freely available to install on your computing machines. Due to open-source, it is available to the public for modification and enhancement.

1.4.1.2 Interpreted Language

Python is an interpreted language because the Python interpreter translates Python code into intermediate code and it executes code logic directly, line by line, without the need for a separate compilation step.

1.4.1.3 Platform Independent

Python programs are designed to be platform-independent, meaning that no matter which operating system you use to develop your code, it can be executed on any other machine that has Python.

1.4.1.4 Easy to Use

Python is widely regarded as easy to use due to its simple and readable syntax, which not only closely resembles human language (like other high-level programming languages, i.e., C++, C#, Java, etc.) but has clear rules about good coding practices through indentation that ultimately leads it closer to the English language.

1.4.1.5 Versatile

Python can be used for a wide range of applications, i.e., web development, data analysis, artificial intelligence, and scientific computing. This flexibility allows developers to undertake projects of different types with the same language.

1.4.1.6 Powerful

Python is powerful as it encompasses several key aspects of advanced programming languages. For example, Python

- can be used for a wide range of application development
- supports diverse libraries and frameworks to simplify and organize complex tasks.
- can easily integrate with other programming languages (i.e., C#, C++, Java, etc.) that allows programmers to take advantage of work already done in other languages and also makes it suitable for various environments and systems.

1.4.1.7 Multiprogramming Paradigm Support

Python supports multiple programming paradigms, i.e., procedural, object-oriented, and functional programming paradigms, which allow developers to choose the best approach for their projects.

1.4.1.8 Strong Community Support and Well-Maintained Documentation

Python has a very strong and active community including a wide range of developers (from beginners to experts), mailing lists, and online platforms that are contributing to its growth. Moreover, Python's official documentation is extensive and well-maintained, which is a good source for learning this language.

1.5 Getting Started with Python

Before diving into writing Python programs to solve agricultural problems, you will need to install the Python language and related Integrated Development Environments (IDEs) on your computer. Here is a step-by-step guide for setting up Python and a suitable IDE on a Windows Operating System (OS).

1.5.1 Installing Python

Follow the step-by-step details given below to complete the installation of Python on your system.

Step 1: Go to the official Python website [17] and download the Windows installer of the latest version of Python (i.e., version 3.13.3 at the time of this writing), as shown in Fig. 1.3.

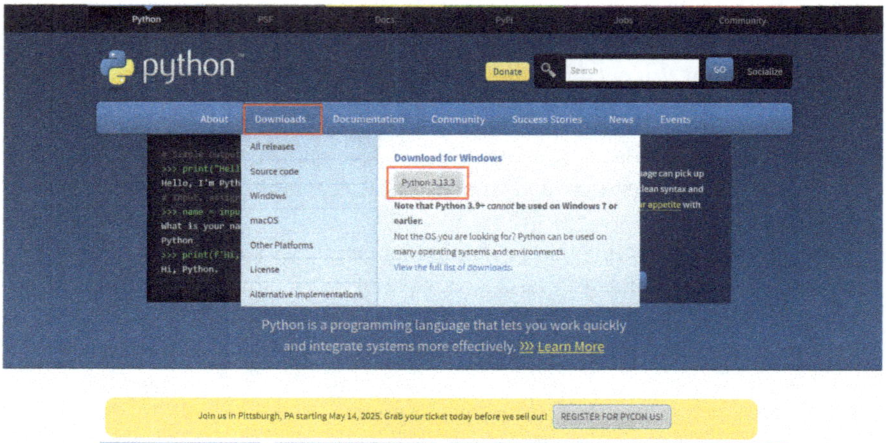

Fig. 1.3 Availability of the latest version of Python Windows installer on the official Python website

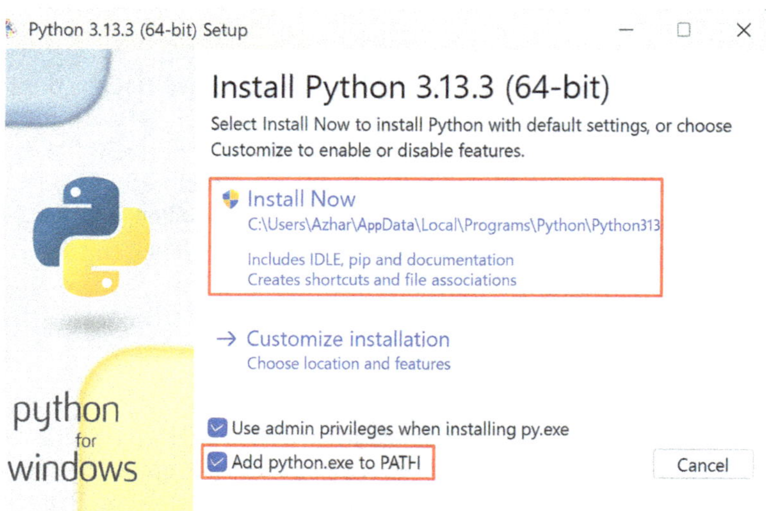

Fig. 1.4 Python installation wizard

Step 2: Run (by double-clicking on) the downloaded Python Windows installer.

Step 3: Check both the boxes of "*Use admin privileges when installing py.exe*" and "*Add python.exe to PATH*" as shown in Fig. 1.4. The checked option "*Use admin privileges when installing py.exe*" allows you to install with admin privileges. Selecting the "*Adding Python to the system's PATH*" option allows you to run Python scripts and access Python packages from any directory in the Command Prompt or terminal.

Step 4: Follow the on-screen instructions and accept the default configurations. Once you are done, you will have Python 3.13.3 on your system.

Step 5: Open the Command Prompt (by pressing *Windows + R*, typing *cmd*, and hitting *Enter*) and type ***python-V*** to verify the successful installation. Executing this command, you will see the Python version installed on your device (Python 3.13.3 in this case, as shown in Fig. 1.5).

1.5.2 Launching Python in Command Prompt

To launch Python in the Command Prompt, open the Command Prompt (by pressing *Windows + R*, typing *cmd*, and hitting *Enter*) and then type *python* or *python3* and press *Enter* (as shown in Fig. 1.6).

Note: Write exit or quit() and press Enter to exit from Python mode in Command Prompt.

1.5.2.1 Writing/Running Python Statements in Command Prompt

After launching Python in the Command Prompt, you can write Python statements for line-by-line execution as shown in Fig. 1.7.

It is important to note that this method is not designed to write and execute multiple Python statements interactively in one go. However, it is suitable for executing Python scripts.

Fig. 1.5 Verification of successful Python installation using MS Windows command prompt

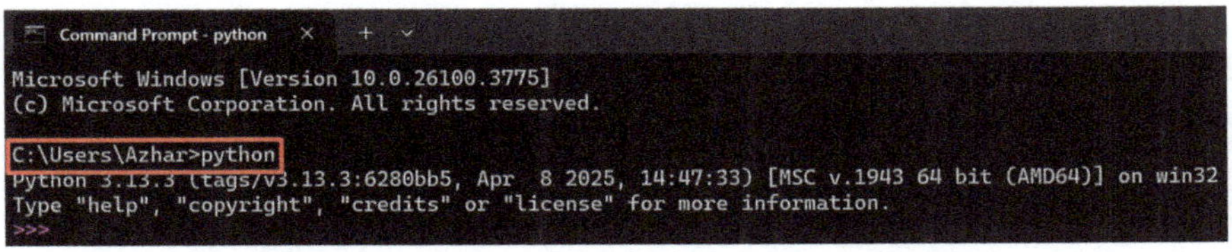

Fig. 1.6 Launching Python in command prompt

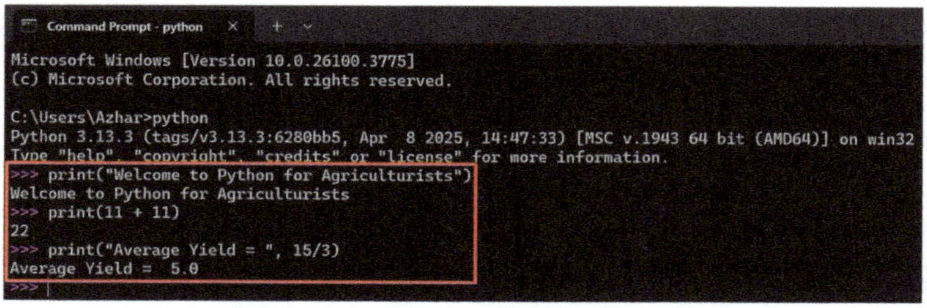

Fig. 1.7 Writing/executing Python statements in command prompt

Fig. 1.8 Executing Python script in command prompt

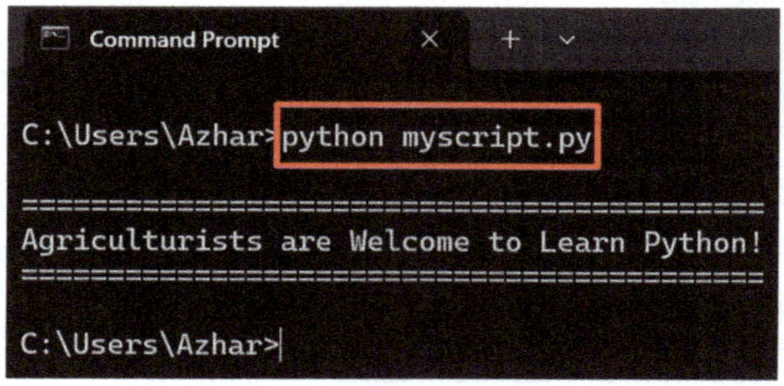

1.5.2.2 Executing Python Scripts using Command Prompt

Suppose a Python script named *myscript.py* containing the following three commands is saved at *C:\Users\Azhar*.

```
myscript.py
print("===========================================")
print("Agriculturists are Welcome to Learn Python!")
print("===========================================")
```

To run this file from the Command Prompt, first navigate to the script's directory using the change directory command (*cd* with path where file is stored) and then execute the script with the *python myscript.py* command, as shown in Fig. 1.8.

1.5.3 IDEs for Python Programming

An IDE (Integrated Development Environment) is an application that typically includes a code editor, compiler or interpreter, and debugger to help programmers write, compile, and debug code. However, additional features of IDEs vary and can include syntax highlighting, intelligent code completion, refactoring support, form designing, and split screen modes that make the overall software development easier and more efficient. A brief description of these typical features or components of an IDE has been given as follows:

- *Code editor* assists programmers in writing/editing code, often with features such as auto-correct, auto-suggestion, automatic line numbering, color coding, etc.
- *Compiler or Interpreter* that translates the source code into a computer-understandable language.
- *Debugger* helps programmers to identify/fix errors through variable inspection, setting breakpoints, and stepping through code.
- *Syntax highlighting* shows language keywords, structs, syntax errors, and so on in different colors and font effects.
- *Intelligent code completion* suggests specific lines of code to avoid common mistakes.
- *Refactoring support* allows system-wide code changes without affecting overall program behavior.
- *Form designing* facilitates programmers to use a drag-and-drop interface to build forms.
- *Split screen mode* supports the editing of multiple source files at once.

A number of IDEs have been developed to facilitate Python programmers. The salient features of five popular Python IDEs have been briefly described in Table 1.3. This chapter includes details regarding the use of IDLE, VS Code, and PyCharm IDEs for Python code writing and execution. However, in the rest of the chapters, all example code snippets are executed in PyCharm.

1.5.3.1 IDLE

The IDLE is a Free and default editor that accompanies Python and can be used on Windows, Linux, and Mac operating systems. Below are the steps to launch and use IDLE for Python on the Windows OS.

Step 1: On the Windows OS, in the *Start Menu* search and select the IDLE installed on your computer. Upon selecting, the IDLE shell will open in interactive mode, as shown in Fig. 1.9.

Step 2: You can type and run Python code line by line directly here in this shell, as shown in Fig. 1.10.

Step 3: To write the multiline script, open the new editor window by clicking on *File* → *New File* in the IDLE shell menu. Write your code here and save the work (using *Ctrl* + *S*) with the .py extension (shown in Fig. 1.11).

Step 4: To execute this saved multiline script, go to *Run* → *Run Module* in the IDLE shell menu. The output of the script will appear in a new window as shown in Fig. 1.12.

Table 1.3 Popular IDEs' features

IDE name	Features
IDLE (Integrated Development and Learning Environment)	– Suitable for beginner-level developers – Smart indenting, along with basic text editor features – Interactive interpreter with syntax highlighting, error, and I/O messages – Have an efficient debugger
Jupyter	– Free and widely used in the field of data science – Interactive and allows live code sharing and visualization – Integration of data science libraries such as NumPy, Pandas, and Matplotlib
PyDev	– Free Python IDE for Eclipse IDE – Provide good support for Python web development – Django framework integration – Supports refactoring, type hinting, and code analysis
Visual Studio Code (VS Code)	– Have enhanced editing tools, including multi-cursor selection, column selection, outline view, side-by-side preview, and search and modify – Support Git integration to collaborate on code – Support code linting – Allow Python programmers to set breakpoints, inspect variables, use the debug console, and configure test environments to run and debug tests
PyCharm	– Suitable for professional developers and facilitates the development of large Python projects – Smart indenting, along with basic text editor features – Interactive interpreter with syntax highlighting, error, and I/O messages – Efficient debugger with persistent breakpoints and stepping

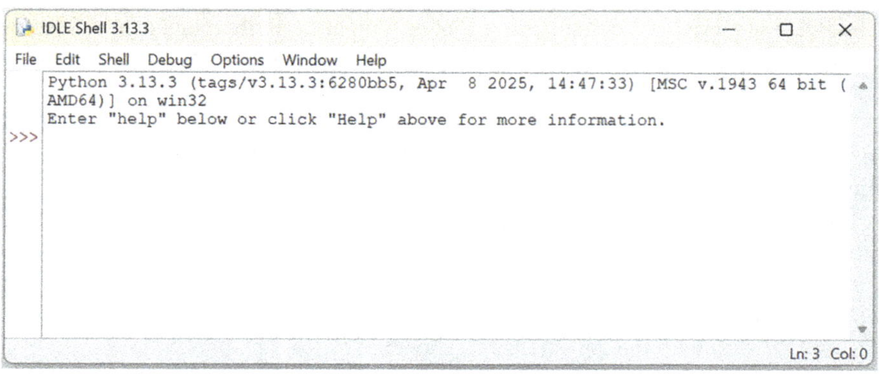

Fig. 1.9 The IDLE shell

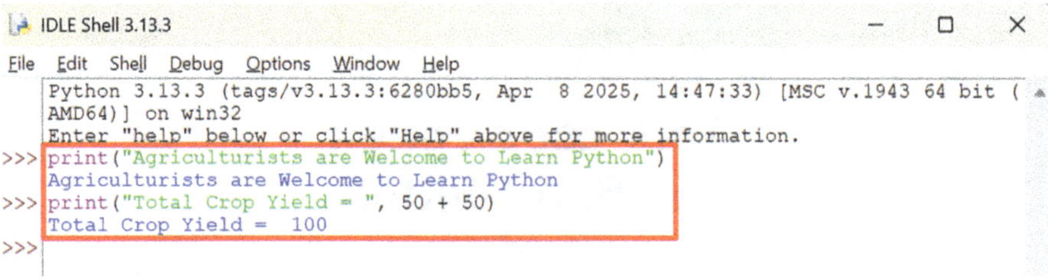

Fig. 1.10 Line-by-line code writing and execution in IDLE

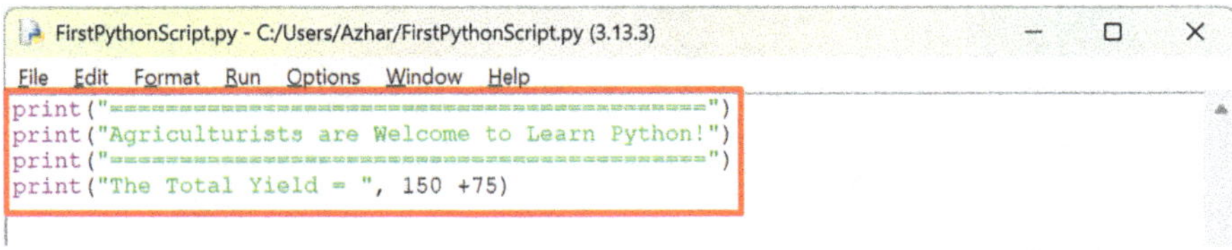

Fig. 1.11 Multiline script in IDLE

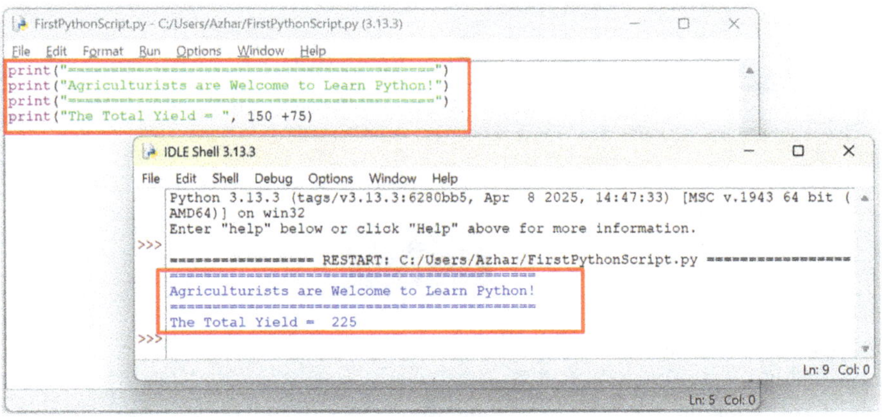

Fig. 1.12 Output of multiline script in IDLE

1.5.3.2 VS Code

VS Code is a lightweight, free, open-source platform that is developed by Microsoft and works well on Windows platforms. Follow the steps given below to download, install, launch, and use VS Code to write and execute Python programs.

Step 1: Download the VS Code Windows installer from the official Visual Studio website [18], as shown in Fig. 1.13.

Step 2: Run the downloaded installer by double-clicking on it and follow the installation wizard (i.e., accept license terms, choose installation location, select options such as "*Add to PATH*" and "*Register as code editor*"). After these

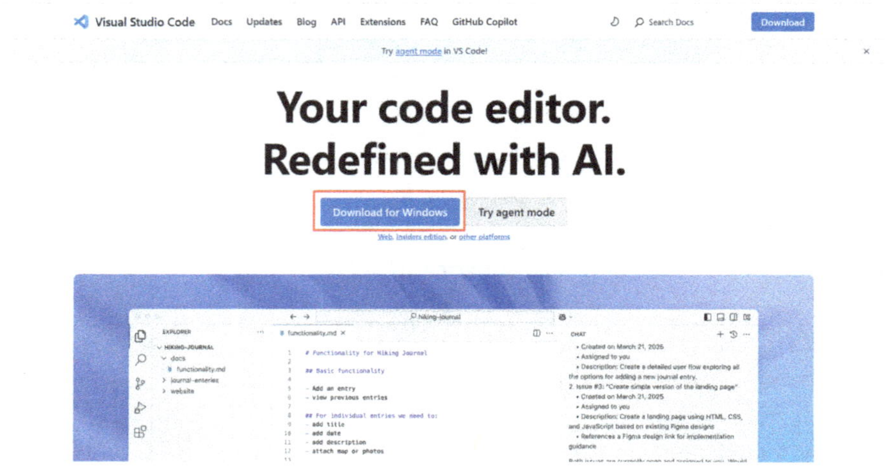

Fig. 1.13 Official website to download VS Code

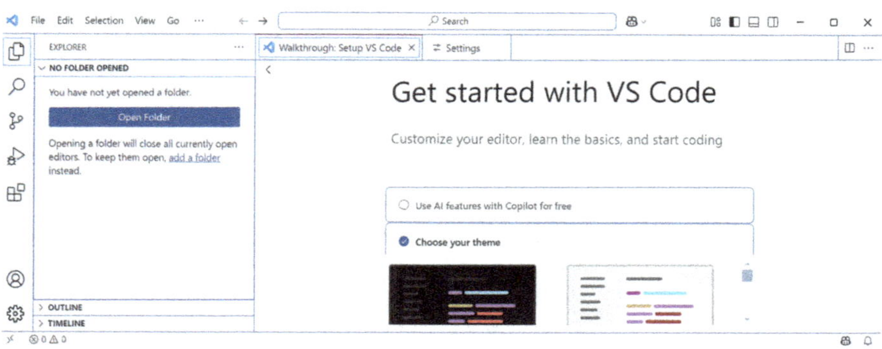

Fig. 1.14 VS Code window

selections, click the "*Install*" button and press the "*Finish*" button after successful installation on your system.

Step 3: After launching VS Code from the Windows Start Menu or an already created VS Code Desktop Shortcut, VS Code will open in interactive mode as shown in Fig. 1.14.

Step 3: Go to the *Extensions* panel (by pressing *Ctrl+Shift+X*) and search for Python and install the extension by Microsoft as shown in Fig. 1.15.

Step 4: Assuming Python is already installed on your system, the next step is to Configure Python Interpreter (by pressing *Ctrl+Shift+P* and typing *Python: Select Interpreter*, and selecting the correct Python Version), as shown in Fig. 1.16.

Step 5: Write the Python code in the VS Code editor and execute by clicking the play icon in the top right and selecting "*Run Python File in Dedicated Terminal*". The output of the code will appear in the terminal as shown in Fig. 1.17.

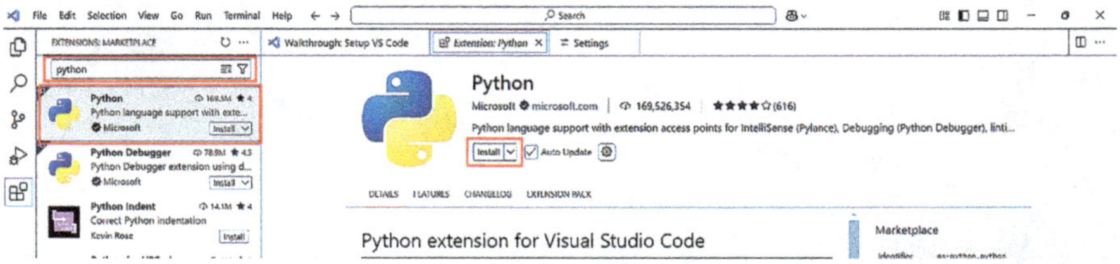

Fig. 1.15 Python extension installation for VS Code

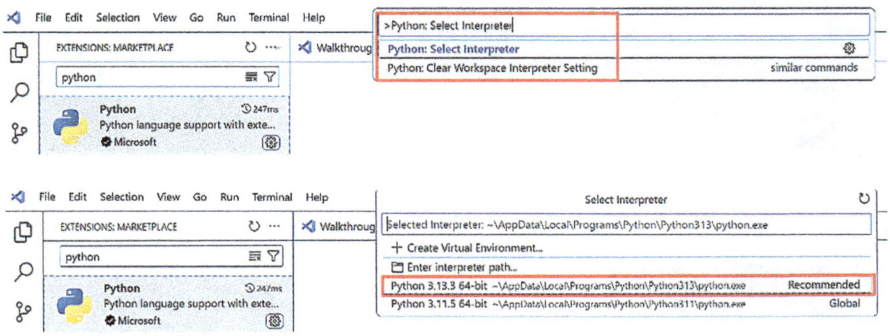

Fig. 1.16 Python configuration in VS Code

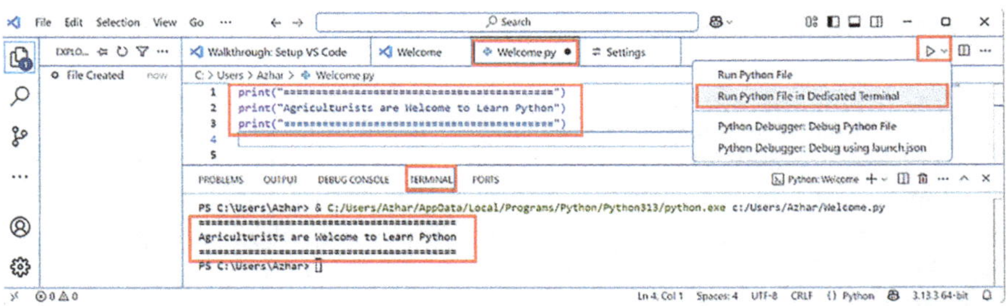

Fig. 1.17 Python code writing and execution in VS Code IDE

1.5.3.3 PyCharm

PyCharm is a Freemium IDE created by JetBrains and can be used on Windows, Mac, and Linux OS. Below are the steps to install, launch, and use PyCharm for Python on Windows OS.

Step 1: Download the PyCharm Windows installer from the official JetBrains website [19], as shown in Fig. 1.18.

Step 2: Run the downloaded installer by double-clicking on it and follow the installation wizard (i.e., selecting options for "*Create Desktop Shortcut*" and "*Add to the PATH*"), click the "*Install*" button, wait for installation to be completed, and then press the "*Finish*" button at the end.

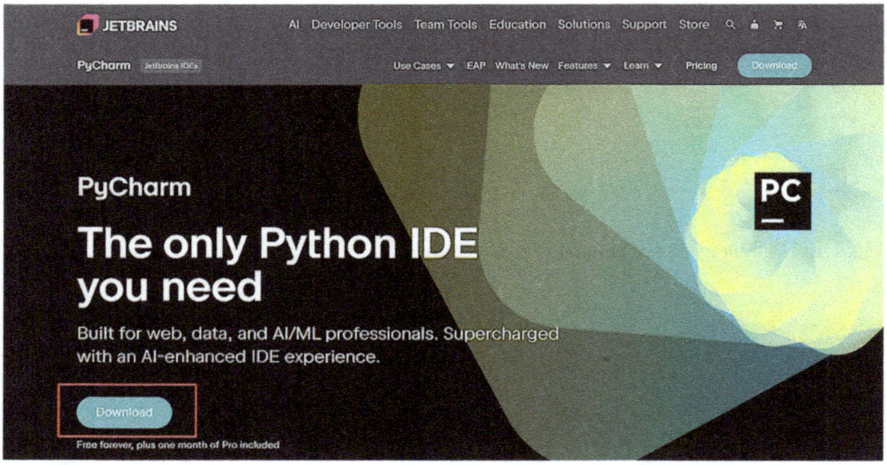

Fig. 1.18 Official JetBrains webpage to download PyCharm

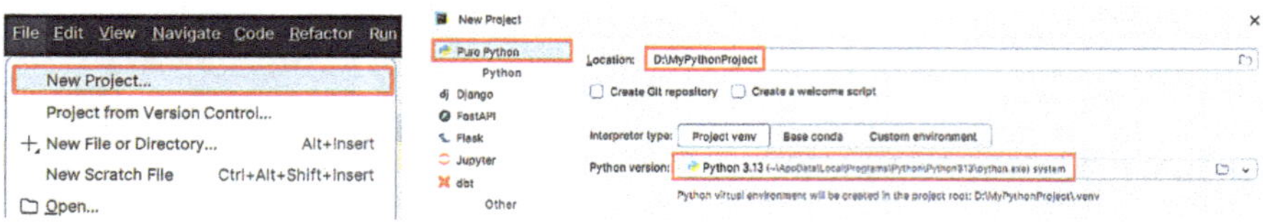

Fig. 1.19 New Project creation in PyCharm

Step 3: After rebooting your computer, you can launch PyCharm from the Start Menu or by double-clicking its desktop shortcut. When opening PyCharm for the first time, it will run in interactive mode. To begin coding in Python, create a new project by navigating to the *File* menu and selecting *New Project*... option. In the New Project window, specify the project name and its location, then choose the appropriate Python interpreter version as shown in Fig. 1.19. This setup ensures that your Python environment is correctly configured for development.

Step 4: To create a Python file in your newly created project shown in the folder pane, right-click on the project name, go to *New*, and then select *Python File* option. A dialog will appear where it is required from you to provide the desired file name. Pressing *Enter* after providing the file name creates a new .py file, as illustrated in Fig. 1.20.

Step 5: In this newly created Python file, you can start writing your Python code in the PyCharm Editor. To execute this code, click on the *play icon* at the top bar, and the output will be displayed in the terminal (the bottom pane), as illustrated in Fig. 1.21.

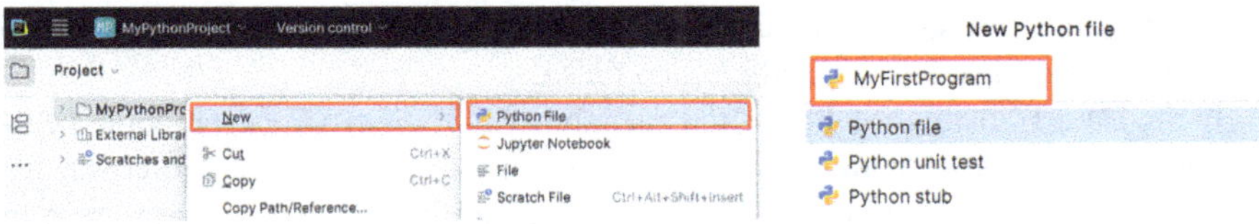

Fig. 1.20 New Python file creation in PyCharm

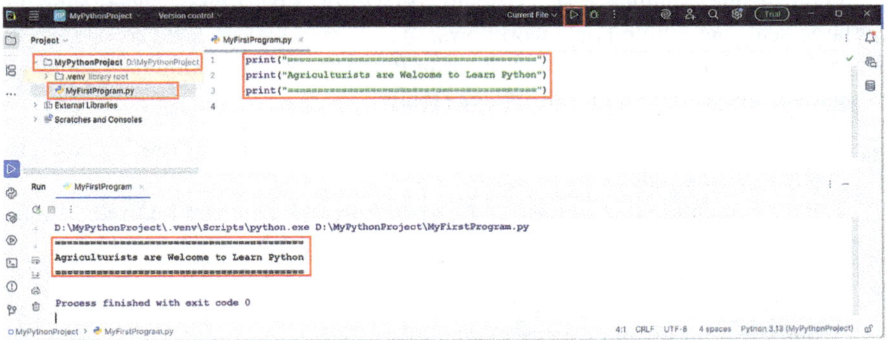

Fig. 1.21 Writing and executing a Python File in PyCharm

1.6 Relationship Between Digital Agriculture Ecosystem and Python

Understanding the close integration between modern agricultural practices and Python programming is essential. Python serves as a powerful tool throughout all phases (i.e., data acquisition, transmission, storage, and analysis, as discussed in Sect. 1.1.3 of this chapter) of data processing in a Digital Agriculture Ecosystem. Mainly, it lays the groundwork for applying data science techniques, such as exploratory data analysis, statistical and mathematical modeling, machine learning, deep learning, and data visualization [20]. These capabilities of the Python language enable agricultural stakeholders to make informed decisions by extracting meaningful and actionable insights from available agricultural data. Below is a brief description of how Python is playing its role in the domain of Digital Agriculture.

- In the Data Acquisition Phase, Python programming assists in the gathering of raw data from both hardware (i.e., sensors and IoT devices) and digital sources (i.e., data stored in structured, semi-structured, and unstructured file formats). In this regard, several Python packages and libraries, i.e., *PySerial*, *Paho-MQTT*, and *Socket* are used to interact with sensors and IoT devices that ultimately enable the real-time collection of environmental data, i.e., soil moisture, temperature, humidity, pH value, and so on. In addition, Python's other packages and libraries (such as *NumPy*, *Pandas*, *PyMySQL*, *BeautifulSoup*, *OpenCv-Python*, etc.) assist in the extraction of data from structured data sources (such as SQL databases), semi-structured data files (such as CSV, XML, JSON, etc.), and unstructured files (such as audio/video files).

- In the Data Transmission Phase, the availability of Python's *Socket* module facilitates network-level communication and reliable data exchange. Moreover, for higher-level data integration, libraries such as the Python *DLT* library help to collect and standardize data from diverse sources. In similar ways, the Python Apache Airflow framework is especially effective for managing data pipelines to automate the complex ETL (Extract, Transform, Load) tasks and organizing large-scale agricultural data workflows.
- In the Data Storage Phase, the following Python tools and techniques are available to handle different needs of data storage.
  - Python *file handling mechanism* with built-in functions for reading, writing, and managing files.
  - Python compression libraries, i.e., *gzip*, *bz2*, and *lzma* can be used to compress data for efficient storage.
  - Python's *psycopg2* library and *Snowflake Connector* can be used not only to connect Amazon Redshift and Snowflake data warehouses but also for executing SQL queries and data manipulation operations.
- In the Data Analytics Phase, a number of Python packages and libraries (i.e., *NumPy*, *Pandas*, *Matplotlib*, *Seaborn*, *Scikit-learn*, *Tensorflow*, etc.) not only provide foundational support for manipulation of numerical and tabular data but also support high-level data visualization and machine-learning tasks for prediction and decision-making.

1.7 Exercises

Problem 1.1 Using Command Prompt, execute a Python script named *MyPythonScript.py* that is saved in a folder named *Python_Programs* on the *D Drive* of your computer hard drive and contains the following code lines:

```
print("================================================")
print("It's Fun to apply Python in Digital Agriculture!")
print("================================================")
print("It's Fun to apply Python in Smart Agriculture!")
print("================================================")
print("It's Fun to apply Python in Precision Agriculture!")
```

Problem 1.2 Open IDLE code editor and write a Python program that prints the following:

```
*****************
***Agriculture***
*       @       *
***   Python   ***
*****************
```

Problem 1.3 Open VS Code and write and execute the following multiline script to check the output in the terminal:

```
# Information about an Agriculture Field
yield_in_summer = 20 # in kg
yield_in_winter = 35 # in kg

profit_per_kg = 3 # in $

spray_cost = 20 # in $
fertilizer_cost = 30 # in $

# total yield
total_yield = yield_in_summer + yield_in_winter

# total cost
total_cost = spray_cost + fertilizer_cost

print("The Total Cost = ", total_cost)
print("Total Yield = ", total_yield)
print("Total Profit = ", total_yield * profit_per_kg)
```

Problem 1.4 Launch PyCharm and create a New Project. Following that, create three Python files in that project and execute each file one by one to display the following:

```
Output of First Python File
===============================
Welcome to Agriculture!
===============================

Output of Second Python File
@@@@@@@@@@@@@@@@@@@@@@@@@@@@@@@
Welcome to Python!
@@@@@@@@@@@@@@@@@@@@@@@@@@@@@@@
Output of Third Python File
^^^^^^^^^^^^^^^^^^^
Welcome to PyCharm!
^^^^^^^^^^^^^^^^^^^
```

References

1. M. A. Iqbal, *Digital Agriculture*. Springer, 2024.
2. D. R. Harris and D. Q. Fuller, "Agriculture: definition and overview," *Encyclopedia of global archaeology,* pp. 104-113, 2014.
3. B. Chandrasekaran, K. Annadurai, and E. Somasundaram, *A textbook of agronomy*. New Age International Limited, 2010.
4. K. Huang *et al.*, "Photovoltaic Agricultural Internet of Things Towards Realizing the Next Generation of Smart Farming," *IEEE Access,* vol. 8, pp. 76300-76312, 2020.
5. M. Ayaz, M. Ammad-Uddin, Z. Sharif, A. Mansour, and E.-H. M. Aggoune, "Internet-of-Things (IoT)-Based Smart Agriculture: Toward Making the Fields Talk," *IEEE Access,* vol. 7, pp. 129551-129583, 2019.
6. "German Agricultural Society (Deutsche Landwirtschafts-Gesellschaft (DLG))," https://www.dlg.org/en/.
7. W. H. Griepentrog, N. Uppenkamp, and R. Horner, "Digital Agriculture , A DLG Position Paper," *DLG e.V., Frankfurt.,* 2018.
8. "National Geographic - Ecosystem Definition." (accessed.
9. V. Udutalapally, S. P. Mohanty, V. Pallagani, and V. Khandelwal, "sCrop: A Internet-of-Agro-Things (IoAT) Enabled Solar Powered Smart Device for Automatic Plant Disease Prediction," *arXiv preprint arXiv:2005.06342,* 2020.
10. A. Al-Fuqaha, M. Guizani, M. Mohammadi, M. Aledhari, and M. Ayyash, "Internet of things: A survey on enabling technologies, protocols, and applications," *IEEE communications surveys & tutorials,* vol. 17, no. 4, pp. 2347-2376, 2015.
11. O. Elijah, T. A. Rahman, I. Orikumhi, C. Y. Leow, and M. N. Hindia, "An overview of Internet of Things (IoT) and data analytics in agriculture: Benefits and challenges," *IEEE Internet of Things Journal,* vol. 5, no. 5, pp. 3758-3773, 2018.
12. S. Wolfert, L. Ge, C. Verdouw, and M.-J. Bogaardt, "Big data in smart farming–a review," *Agricultural Systems,* vol. 153, pp. 69-80, 2017.
13. T. Daum, "Digitalization and skills in agriculture," *Outlook on Agriculture,* p. 00307270251336474, 2025.
14. H. B. Mehare, J. P. Anilkumar, and N. A. Usmani, "The Python programming language," in *A guide to applied machine learning for biologists*: Springer, 2023, pp. 27-60.
15. V. K. Sharma, V. Kumar, S. Sharma, and S. Pathak, *Python Programming: A Practical Approach*. Chapman and Hall/CRC, 2021.
16. D. Xanthidis, C. Manolas, O. K. Xanthidou, and H.-I. Wang, *Handbook of Computer Programming with Python*. CRC Press, 2022.
17. "Python," https://www.python.org/.
18. "VS Code," https://code.visualstudio.com/.
19. "PyCharm," https://www.jetbrains.com/pycharm/download.
20. S. Jha, J. V. Krogmeier, D. R. Buckmaster, and A. D. Balmos, "Python Programming in Digital Agriculture," in *Case Studies and Modules for Data Science Instruction*: American Society of Agricultural and Biological Engineers, 2020, pp. 7-24.

Chapter 2
Basic Structure and Elementary Components of Python Programs

2.1 Introduction

Python programming components or constructs can be divided into several categories, i.e., Elementary Programming Constructs, Control Flow Structures (Selection and Iteration), Functions (Built-in and Custom), Modules (Built-in and Custom), Packages (Built-in and Custom), and Libraries (Built-in and Custom). Familiarity with these components is essential to create Python programs ranging from simple code scripts to complex software systems. Hence, the focus of this chapter is to introduce you to writing and executing elementary Python programs to solve numerical problems in different domains of agriculture. By understanding the anatomy of these small programs (discussed in this chapter), you will learn the main steps of effective Python program writing style using Python Elementary Constructs, including the use of comments, variables, constants, arithmetic operators, solving expressions, getting user input, displaying calculated results, and error/exception handling.

2.2 Python Statements and Programming Style

A well-written Python program consists of two types of statements, i.e., non-executable statements (also known as comments) and executable statements. To understand the Python programming style, including effective use of comments along with the appropriate spacing and correct indentation of Python statements, consider the implementation (Listing 2.1) of a simple agriculture problem (Problem 2.1).

M. A. Iqbal, *Python for Agriculturists*,
https://doi.org/10.1007/978-3-032-01442-9_2

Problem 2.1 A farmer expects a yield of 2000 kg of maize per hectare, and the market price for maize is US$ 0.50 per kg. Production costs are US$ 300 per hectare. Write a program to calculate the profit from yield for a 10-hectare field. The formulae to calculate the crop yield profit are as follows:

$$\text{Profit} = \text{Total Revenue} - \text{Total Cost}$$

$$\text{Total Cost} = \text{ProductionCost} / \text{Hectare} \times \text{Total Area}$$

$$\text{Total Revenue} = \text{Total Yield} \times \text{MarketPrice} / \text{kg}$$

$$\text{Total Yield} = \text{Yield} / \text{Hectare} \times \text{Total Area}$$

Listing 2.1	Program to calculate crop yield profit

```
# Python program to calculate the crop yield profit by
# considering various factors such as yield/hectare,
# market price, per unit production cost to determine
# the profitability of an agriculture field.

# Variable declaration and Initialization
yield_per_hectare = 2000                # kg
market_price_per_kg = 0.50              # dollars
production_cost_per_hectare = 300       # dollars
field_size = 10                        # hectares

# Calculate total yield
total_yield = yield_per_hectare * field_size

# Calculate total revenue
total_revenue = total_yield * market_price_per_kg

# Calculate total production cost
total_cost = production_cost_per_hectare * field_size

# Calculate profit
profit = total_revenue - total_cost

# Output the profit
print(profit)
```

In Listing 2.1, you can see the use of comments (statements beginning with the # symbol) and non-executable statements (written without the # symbol). The details regarding the appropriate use of Python comments, along with the structure and types of executable statements, are provided below.

2.2.1 Appropriate Comments

In Python, comment statement(s) start with the # symbol and these comments (like other high-level computer programming languages) are just notes (non-executable statements) that are added to the source code and are helpful for programmers (and other people) to understand the semantics of the executable program statement(s). In Listing 2.1, lines 1–4, 5, 10, 12, 14, 16, and 18 are all examples of (single-line) comments. Some comments are in-line comments as these are placed within the executable code line. For example, comments at lines 6–9 of Listing 2.1 are in-line comments. You can see that all these comments are human-readable annotations in a program's source code that make it easier to understand the working of executable statement(s). For example, the comments at lines 1–4 describe the purpose of this program, and the in-line comments at lines 6–9 describe units of yield, price, production cost, and field area, and so on. Similarly, to wrap up, the key takeaway on comments is as follows:

- keep comments clear and brief to avoid overcrowding and ensure readability, making the code structure clear and manageable.
- it is good practice to begin your program with concise summary comments explaining the purpose and key features of the program.
- for big computer programs, it is recommended to add comments to introduce each main section and clarify complex parts.
- comments are non-executable program statements and do not affect the output of the program (therefore, in Listing 2.1, the deletion of lines 1–4, 5, 10, 12, 14, 16, and 18 and in-line comments at lines 6–9 will not affect the program execution and output).

Mostly in programming languages, in addition to in-line and single-line comments, there are also multiline (or paragraph) comments. Technically, Python does not have a syntax for multiline comments, and programmers insert the # symbol for each code line. However, to facilitate programmers, Python allows programmers to use triple (single or double) quotes for multi-line comments. For example, the first four lines of Listing 2.1 can be written as

```
'''
Python program to calculate the crop yield profit by
considering various factors such as yield/hectare,
market price, per unit production cost to determine
the profitability of an agricultural field.
'''
```

Or

```
"""
Python program to calculate the crop yield profit by
considering various factors such as yield/hectare,
market price, per unit production cost to determine
the profitability of an agricultural field.
"""
```

2.2.2 Well-Structured Python Executable Statements

The executable statements are the instructions written in the source code and are executed line by line from top to bottom to provide the desired output. In simple words, executable statements serve as the building blocks of a computer program. Python programs are composed of a sequence of one or more executable statements. Lines 6–9, 11, 13, 15, 17, and 19 are executable statements in Listing 2.1. It is advisable to consider the following guidelines for writing executable statements in Python.

Use indentation consistently: Indentation involves the use of white spaces (through the tab key or spacebar key) at the beginning of lines of code. Unlike other programming languages where indentation in code is for readability purposes only, in Python, it is essential for defining code blocks. Therefore, proper indentation serves as both a visual separation (to improve code readability) and a structural requirement (to define code blocks clearly, indicating where code block begins and ends) in Python programs. In Python, it is an error if the program statements are not properly indented.

Separate logical section: Use blank lines to separate statements that are logically related to each other (representing logical sections in a computer program). For example, see blank lines without numbering in Listing 2.1.

Space with binary operators: Adding space on both sides of any binary operator (e.g., +, –, /, *, etc.) (details are available in Sect. 2.3.4) is also recommended. For example, see Lines 6–9, 11, 13, 15, 17, and 19 for the use of spaces on both sides of "=" operator or Lines 11, 13, 15, 17 and 19 for the use of spaces on both sides of "=", "*", and "–" operators.

Keep code lines short: As long code lines can make code harder to read, it is recommended to limit code line length to 79 characters.

The most basic components of Python statements are discussed in detail in the next section under the heading of Elementary Components.

2.3 Elementary Components

This section elaborates the use of basic elementary components of the Python programming language. These components are actually the elements of Python's executable statements.

2.3.1 *Identifiers*

Identifiers are the names that can be given to variables, constants, functions, methods, classes, objects, modules, packages, and so on. Rules for naming a Python identifier are given as follows:

- An identifier can be a sequence of letters, digits, and underscores (_), but it must not start with a digit. By convention, the identifier should start with a letter rather than an underscore.
- White spaces and special symbols (#, &, %, $, !, @, etc.) are not allowed.
- Identifiers are case-sensitive and can be of any length.
- Descriptive identifiers are recommended that make programs easy to read, and therefore, it is better to avoid using abbreviations for identifiers.
- It cannot be a Python keyword. (The list of Python keywords along with a brief description is shown in Table 2.1.)

Note: As of Python 3.13, there are 36 keywords in Python (shown in Table 2.1) that cannot be used as identifiers. There can be more keywords in future versions of Python and therefore to see the latest list of Python keywords, you can use the ***print(keyword.kwlist)*** *statement. Moreover, it is important to note that all keywords are case-sensitive. For example, the keyword import cannot be written as Import, IMPORT, ImPort, etc.*

2.3.2 *Variable, Constant, and Data Types*

2.3.2.1 Variable

Variables are containers that can store literal data values. In Python, unlike some other high-level programming languages, variables do not need a separate declaration statement; they are created automatically when assigned a data value for the first time in the Python program. Variables in Python are mutable, which means their values can be changed after their creation. In Listing 2.1, *yield_per_hectare*, *market_price_per_kg*, *production_cost_per_hectare*, and *field_size* are examples of variables to store (literal) data values of *2000*, *0.50*, *300*, and *10*, respectively. Moreover, *total_yield*, *total_revenue*, and *total_cost* are also variables that can store the value of calculated results.

Table 2.1 Python keywords

No.	Keyword	Description
1	import	Includes a module in the current Python code/script
2	from	Used with import keyword to import specific parts of a module
3	as	Used to create an alias, often with imports (e.g., import module as alias)
4	None	Represents a null value or absence of a value
5	True	Boolean value representing truth (logical 1)
6	False	Boolean value representing false (logical 0)
7	and	Logical (binary) AND operator that returns true if both operands are true
8	or	Logical (binary) OR operator that returns true if any operand is true
9	not	Logical (unary) NOT operator that returns true if the operand is false
10	if	Starts a conditional block
11	elif	Conditional statement that is shorthand for "else if"
12	else	Specifies code to run if the preceding if or elif condition is not met
13	while	Used to start a loop that continues as long as a condition is true
14	for	Used to start a loop that iterates over a sequence
15	break	Used to exit the closest enclosing loop prematurely
16	continue	When used it skips the rest of the current loop iteration and moves to the next iteration
17	def	Used to define a function or method
18	return	Exits a function (or method) and optionally returns a value
19	lambda	Defines an anonymous (in-line) function
20	async	Used to define asynchronous functions with async def
21	await	Used to pause and resume asynchronous functions
22	try	Specifies a block of code to test for errors
23	except	Specifies how to handle exceptions in a try block
24	finally	Specifies a block of code that must execute after try and except
25	with	Simplifies exception handling with context managers (e.g., file handling)
26	class	Used to define a class in Python
27	is	Tests object identity; checks if two objects refer to the same memory location
28	pass	A placeholder statement that does nothing; used for unimplemented code
29	raise	Keyword used to trigger an exception
30	assert	Debugging tool to test conditions and it raises an error if the condition is false
31	yield	Used in generator functions to return a value and pause execution
32	in	Tests membership in a collection or sequence
33	del	Deletes a reference to an object
34	nonlocal	Declares a variable from an enclosing (non-global) scope to be used in the current scope
35	global	Declares a variable as global, allowing modification of variables outside the current scope
36	case	Used in the context of structural pattern matching

2.3.2.2 Constant

In programming, constants are names associated with values that remain fixed and unchanging throughout the execution of a program. Python does not have a specific syntax to define constants. Instead, constants are created and initialized in the same way as Python variables. Therefore, although rules to define constants are the same as for variables; however, to facilitate programmers, the Python community has a consensus about the naming convention for Python constants that it must include uppercase letters only. In this way, constant(s) in Python programs can be differentiated from variable(s). Enhanced code readability and improved code maintainability are two main advantages of this convention. The use of conversion factors from *acre* and *hectare* to *square kilometer* in Listing 2.2 is an example of writing or declaring Python constants.

Listing 2.2	Program to calculate the area in square km

```
'''
=====================================================
Python program ensuring the accurate land measurement
for agricultural planning in Square Km.
=====================================================
'''
'''
Constants representing the conversion factor from acres and
hectares to Square Km.
'''
ACRE_TO_SQKM = 0.00405
HECTARE_TO_SQKM = 0.01
# Field areas in acres and hectares
area_in_acres = 49.42 # acres
area_in_hectares = 20 # hectares
# Conversion from acre/hectares to Square Km
aarea_sqkm = area_in_acres * ACRE_TO_SQKM
print("The land area is", aarea_sqkm, "Square KM")
harea_sqkm = area_in_hectares * HECTARE_TO_SQKM
print("The land area is", harea_sqkm, "Square KM")
```

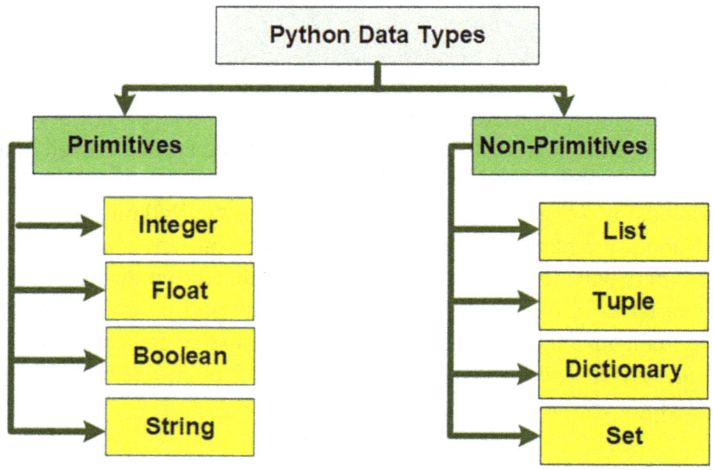

Fig. 2.1 Python datatypes

2.3.2.3 Data Types

The value of a variable or constant in a program can be of any Primitive or Non-Primitive (also known as Compound Data Types or Python Data Structures) data type, as shown in Fig. 2.1. Typically, variables/constants of primitive data types can store one value at a time, and variables/constants of compound data types can store multiple values or multiple pieces of data at a time. A brief description of these data types has been given in Table 2.2. The details and uses of these data types have been discussed in subsequent relevant sections and chapters of this book.

2.3.3 Assignment Statement

In Python, typically an assignment statement (using the assignment operator "=") is used to assign a concrete value and the result of an expression to a variable. The general syntax of an assignment statement is

```
variable_name = value(s) or expression or function_call
```

This syntax shows that an assignment statement has three components, i.e., a left operand that must be a variable or constant, the assignment operator, and a right operand that can be a concrete value (also known as literal such as 0, 3.25Agricultureetc.) or an expression (such as mathematical X + 10/5 + 7, or Boolean expression such as Y > 10, etc.). In Listing 2.2, lines 11 and 12 (related to the declaration and initialization of constants), lines 14 and 15 (related to the

Table 2.2 Python primitive and non-primitive data types

Data type		Description
Primitive	Integer	Used to represent whole numbers from negative infinity to infinity, e.g., 13, 5, −9, 0, etc.
	Float	Stands for "floating point number" and is used for rational numbers, e.g., 1.11, −7.3, 3.14, 13.13, etc.
	Boolean	This data type can hold the values True and False, which are often interchangeable with the integers 1 and 0. Booleans are useful in conditional and comparison expressions. Details of Boolean are discussed in Chap. 3 of this book
	String	It represents the sequence of characters that are enclosed within a pair of single or double quotes, e.g., "cake," "agri@ag.com," "56aST," "786," etc. String details are discussed in Sect. 2.3.6.2 of this chapter
Non-primitive/ compound (or Python Data Structures)	Lists, Tuple, Set, Dictionary	Details of these Python Data Structures have been discussed in Chap. 5

declaration and initialization of variables), and lines 17 and 19 (associated with the use of expression assignments of calculated area in square km) are all examples of assignment statements. Different ways of using assignment statements in Python have been shown below.

2.3.3.1 Basic Assignment

Syntax

```
variable_name = value(s)
variable_name = expression
variable_name = function_call()
```

Examples:

```
# Direct assignment of crop yield in kilograms
crop_yield = 1200 # in kg

# Using an expression to calculate land area
land_area = field_length * field_width

# A function call to get average rainfall of January
average_rainfall = get_average_rainfall("January")
```

2.3.3.2 Multiple Assignments

Syntax

```
variable_name1 = variable_name2 = … variable_nameN = value
variable_name1 = variable_name2 = … variable_nameN = expression
```

Examples:

```
# Assigning the area to agriculture fields of the same size
field1_area = field2_area = field3_area = 50 # Hectares

# Assigning the same fertilizer quantity to multiple fields
field1_fertilizer = field1_fertilizer = available_fertilizer / 2

# Assigning the average winter rainfall value to two months
dec_rainfall = jan_rainfall = get_avg_rainfall("Winter")
```

2.3.3.3 Parallel Assignments

Syntax

```
var_name1, var_name2, …, var_nameN = value1, value2, … valueN
var1, var2, …, varN = expression1,expression2,…,expression
```

Examples:

```
# Assigning crop names in parallel
crop1, crop2, crop3 = "Wheat", "Rice", "Corn"

# Calculating water and fertilizer needed per hectare
water_needed, fertilizer_needed = field1_area * 10, field1_area * 5
```

2.3.3.4 Simultaneous/Parallel Assignments with Iterable Unpacking

Syntax

```
var_name1, var_name2, ..., var_nameN = N_Length_Iterable
(var_name1, var_name2, ..., var_nameN) = N_Length_Iterable
[var_name1, var_name2, ..., var_nameN] = N_length_Iterable
var1, *list_var, ..., varN = Unknown_Length_Iterable
```

Examples:

```
# Assigning values from a list of soil types
soil_types = ["Loamy", "Sandy", "Clayey"]
type1, type2, type3 = soil_types

# Crop yield in kg per hectare
(crop_yield1, crop_yield2, crop_yield3) = (400, 500, 350)

# Assigning different agricultural regions
[region1, region2, region3] = ["North", "South", "East"]
'''
Dal is assigned to first_crop and Oat is assigned to last_crop;
the rest go into mid_crops variable accordingly.
'''
first_crop, *mid_crops, last_crop = ["Dal","Fig","Pea","Oat"]
```

2.3.3.5 Compound or Augmented Assignments

Syntax

```
Existing_variable_name <AR> = value
Existing_variable_name <AR> = expression
```

Note: Here, in the above expressions, <AR> can be any of the Arithmetic *Operators discussed in Sect. 2.3.4.*

Examples:

```
# Adding new harvested crop yield to total yield
total_yield += new_harvest_yield

# Increasing fertilizer requirement by 5%
required_fertilizer *= 0.05
```

2.3.4 Arithmetic Operators and Precedence

From Fig. 2.1, it becomes evident that Python's primitive data types can be divided into text, Boolean, and numeric categories. The two numeric data types are *Integer* and *Float* numbers, which represent numbers without a decimal point *(.)* and numbers with a decimal point, respectively. Table 2.3 shows how Python arithmetic operators can be used to perform arithmetic calculations (in agricultural scenarios).

Table 2.3 Arithmetic operators in Python

Arithmetic operator	Name	Description	Agriculture scenario example	Expression in Python	Result
+	Addition	Adds two values	Yield of two fields (1500 kg + 1200 kg)	1500 + 1200	2700 kg
–	Subtraction	Subtracts one value from another	Expected yield minus losses (2000 kg – 300 kg)	2000 – 300	1700 kg
*	Multiplication	Multiplies two values	Price per kg of crop (US$ 0.5) times total yield (1500 kg)	0.5 * 1500	US$ 750
/	Division	Divides one value by another	Dividing yield over total area (2000 kg/4 ha)	2000/4	500 kg/ha
%	Modulus or Remainder	Finds the remainder	Seeds left after packing (105 seeds % 20 per pack)	105 % 20	5 seeds
**	Exponentiation	Raises a value to the power of another	Doubling yield (2 ** 3 fields to get target yield)	2 ** 3	8 fields
//	Floor Division	Divides and rounds down to nearest integer	Dividing harvest evenly among workers (105 kg//4)	105//4	26 kg

Table 2.4 Arithmetic operator precedence levels

Operator	Precedence level	Evaluation direction for operators
**	Highest (1st)	Left to right
*, /, //, %	Higher (2nd)	Left to right
+, –	Lowest (3rd)	Left to right

In addition to understanding the semantics and usage of arithmetic operators, it is important to know the operator precedence rules of Python. Operator precedence rules determine the order of evaluation in a mathematical expression. Table 2.4 summarizes the arithmetic operator precedence from highest to lowest.

Note: It is important to consider that any expression written in parentheses ( ), braces { }, or brackets [ ] has more precedence over other arithmetic operators in that expression.

2.3.5 *Expression*

An expression in Python is a special type of statement that combines operands (literal(s) and/or variable(s)) and operators to produce a result (as a single value). For example, Line 11 (*total_yield = yield_per_hectare * field_size*) in Listing 2.1 is an expression that evaluates to 20,000 (a value that will be stored in the variable *total_yield* after the multiplication of *yield_per_hectare* value (i.e., 2000) to *field_size* value (i.e., 10)).

2.3.6 *Getting User Input and Displaying Output in Python Programs*

2.3.6.1 Getting Input from the User

In all previous examples in this chapter, variables were assigned data values within the program. However, Python's input statements enable users to input data at execution time. This allows more dynamic interactions and makes the program flexible and adaptable to different values without needing to modify the code.

2.3.6.1.1 For Text Input

The syntax of using the input statement to get text input from the user is

```
variable_name = input()
```

or

```
variable_name = input('Enter your Name')
```

or

```
variable_name = input("Enter your Name")
```

(Note: input() is a function, and we will discuss functions in detail later in Chap. 4.*)*

2.3.6.1.2 For Numerical Input

The syntax of using an input statement to get numerical input (digits or numbers that are used in calculations) from the user is

```
variable_name = eval(input("Enter a Digit or Number"))
```

or

```
variable_name = int(input("Enter a Digit or Number"))
```

or

```
variable_name = float(input("Enter a Digit or Number"))
```

Here, *eval()* is a function that takes a numerical value or mathematical expression as a string and returns the value or result of the expression as an integer or float value. If you want to be more specific in converting to a specific data type, then you can use the *int()* or *float()* functions.

(Note: In case of non-numerical string value, eval(), int(), and float() functions cause an error. Therefore, you must enter a numeric value if some calculations are required to be performed on the given input. Overall, these three functions are useful for manipulating and converting numbers in Python programs.)

2.3.6.2 Displaying Output

To display output, the *print* statement (or the *print()* function) is used in Python, which requires parentheses around its arguments. Different ways to use *print* statements (with and without formatting) have been described below.

2.3.6.2.1 Printing Sequence of Characters (Known as String)

The print() function displays any text (known as a sequence of characters or strings) enclosed in single or double quotes exactly as it is written, as demonstrated below.

```
print("Welcome to Digital Agriculture")
print('Welcome to Digital Agriculture')
```

The output of both print statements will be ***Welcome to Digital Agriculture***.

Certain important aspects of strings (i.e., escape sequences, indexing and slicing), use of separator parameter, mathematical expressions, restricting mathematical output to specific decimal places are relevant to the *print()* statement and worth discussing here.

2.3.6.2.2 Print Special Characters (Use of Escape Sequence)

An escape sequence is a special character sequence that allows the programmer to include special characters (which are not possible to represent directly) within a string. Generally, these are used for formatting text. In Python, typically a backslash character (\) is used for this purpose. The use of escape sequences for some common characters (listed in Table 2.5) is demonstrated in Listing 2.3.

Table 2.5 Common escape sequences in Python

Escape sequence character	Description
\n (New Line or Line Break)	Used to display a new line in a string
\t (Tab character)	Used to show the tab spaces in a string
\” (Double Quote character)	Used to show double quotes in a string
\\ (Backslash character)	Used to show a backslash \ character in a string

Listing 2.3	Program demonstrating the use of escape sequences (mentioned in Table 2.5)

```
# Use of \n to display output on different lines
print("========== Use of \\n ==========")
brief_report = "Crop: Pea\nMoisture: High\nFertilizer Needed: No"
print(brief_report)

# Use of \t to display output in tabulated way
print("========== Use of \t ==========")
crops_data = "Field\tCrop\tYield\n101\tA\t2300kg\n102\tB\t1300kg"
print(crops_data)

# Use of \" to include quotes in the output
print("========== Use of \" ==========")
field_name = "The \"Golden Harvest\" Field"
print(field_name)

# Use \\ for backslash (\) (typical use for file paths)
print("========== Use of Backslash in String ==========")
file_path = "C:\\FarmData\\Harvest2025\\log.txt"
```

Output of Listing 2.3:

The self-explanatory output of Listing 2.3 is shown below.

```
========== Use of Line Break ==========
Crop: Pea
Moisture: High
Fertilizer Needed: No
========== Use of Tab Characters ==========
Field → Crop → Yield
101 → A → 2300kg
102 → B → 1300kg
========== Use of Double Quote ==========
The "Golden Harvest" Field
========== Use of Backslash ==========
C:\FarmData\Harvest2025\log.txt
```

2.3.6.2.3 Print Subset of a Sequence of Characters (Use of Indexing and Slicing)

In Python, strings can be considered as a sequence of indexed characters. To better understand this concept, consider the following line of Python code along with its simplified logical representation in computer memory below.

```
message = "Digital Agriculture"
```

Index	0	1	2	3	4	5	6	7	8	9	10	11	12	13	14	15	16	17	18
Characters	D	i	g	i	t	a	l		A	g	r	i	c	u	l	t	u	r	e

From the above (shown) representation, it becomes evident that the sequence of characters in a String is stored at contiguous memory locations (that can be represented as index values). Therefore, as a programmer, you can easily pick out specific character(s) from this string. For this purpose, indexing and slicing are available in Python. Indexing refers to the process of accessing some specific characters from a given String, whereas slicing refers to the taking of a sequence of characters from one given index to another given index, as demonstrated in Listing 2.4.

Listing 2.4	Program demonstrates the use of *Indexing* and *Slicing*

```
message = "Digital Agriculture"

#Examples of positive indexing
print("First String Character:", message[0])
print("Ninth String Character:", message[8])

# Examples of negative indexing
print("First String Character from Last:", message[-1])
print("Ninth String Character from Last:", message[-9])

# Exmaples of using Slicing
print("First part before space:", message[0:7])
print("Second part after space:", message[8:19])

# Slicing Examples to collect random sequences

# To collect characters from indices 8 to 10
print("Sequence 1", message[8:11])

# To collect characters from indices 0 to 10
print("Sequence 2", message[:11])
```

```
# To collect characters from indices 8 to onwards
print("Sequence 3", message[8:])

# To collect characters from indices -4 to -9
print("Sequence 4", message[-9:-3])

# To collect characters from indices -10 to the start
print("Sequence 5", message[:-9])

# To collect characters from indices -1 to -3
print("Sequence 6", message[-3:])

# To collect characters from index 1 to 9 by twos
print("Sequence 7", message[1:9:2])

# To collect characters in reverse direction from -1 to start
print("Sequence 8", message[ : :-1])
```

Output of Listing 2.4:

The self-explanatory output of Listing 2.4 is shown below.

First String Character: D
Ninth String Character: A
First String Character from Last: e
Ninth String Character from Last: r
First part before space: Digital
Second part after space: Agriculture
Sequence 1 Agr
Sequence 2 Digital Agr
Sequence 3 Agriculture
Sequence 4 ricult
Sequence 5 Digital Ag
Sequence 6 ure
Sequence 7 iia
Sequence 8 erutlucirgA latigiD

Some examples of using indexing and slicing in a smart farming context have been demonstrated in Listing 2.5.

Listing 2.5	Examples demonstrate the use of *Indexing* and *Slicing* in the smart farming context

```
print("=====Example 1=====")
# To check sensor ID for a particular agriculture field
sensor_id = "Field1_SID945"
if sensor_id[0] == "S":
    print("Proper sensor in Field1")
else:
    print("This sensor should be deployed to some other field")
```

```
print("=====Example 2=====")
# To extract date in terms of year, month, and day from a record
yield_record_date = "20231104_log"
year = yield_record_date[0:4]
month = yield_record_date[4:6]
day = yield_record_date[6:8]
print("The record was collected on", day, month, year)

print("=====Example 3=====")
'''
To get a specific location in a specific farm from a coded Farm
ID. Suppose the first three characters represent the field id and
the fourth character represents a specific area in that field e.g.,
S for South area and W for West area
'''
farm_ID = "FD1S25X9"
field_number = farm_ID[0:3]
area_loc = farm_ID[3]
if(field_number == "FD1") and (area_loc == "S"):
    print("This is south part of Field 1 at Farm 1")
else:
    print("This is Not south part of Field 1 at Farm 1")

print("=====Example 4=====")
'''
To determine the nature of reading. For example, if characters 5
and 6 are WS in a data reading then it shows that this data is
collected from a weather station
'''
reading = "Temp-WS-20251150_30"
if(reading[5:7] == "WS"):
    print("Data collected from Weather Station")
else:
    print("Data collected from Sensor")

print("=====Example 5=====")
# Extract Zone, Plot, and Batch information in a single identifier
field_id = "ZoneA_Plot05_Batch3"
zone = field_id[0:5]
plot_number = field_id[6:11]
batch_number = field_id[13:]
print("Crop is planted in", zone, plot_number, batch_number)

print("=====Example 6=====")
# Extract crop code and growth stage from given string
crop_info = "WHT_Mature_WestField"
crop_code = crop_info[0:3]
growth_stage = crop_info[4:10]
```

```
field_location = crop_info[11:]
print("Crop Code:", crop_code)
print("Growth Stage:", growth_stage)
print("Location:", field_location)
```

Output of Listing 2.5:

The self-explanatory output of Listing 2.5 is shown below.

```
=====Example 1=====
This sensor should be deployed to some other field
=====Example 2=====
The record was collected on 04 11 2023
=====Example 3=====
This is south part of Field 1 at Farm 1
=====Example 4=====
Data collected from Weather Station
=====Example 5=====
Crop is planted in ZoneA Plot0 Batch3
=====Example 6=====
Crop Code: WHT
Growth Stage: Mature
Location: WestField
```

2.3.6.2.4 Print with Separator Parameter

Python by default inserts a space between each of the arguments (separated by a comma) of the print function. However, using an optional argument called *sep* (short for separator), that space can be changed to something else. See the output of the following three print statements

```
print("Bean", "Pea", "Oat") → Output: Bean Pea Oat
print("Bean","Pea","Oat",sep="") → Output: BeanPeaOat
print("Bean", "Pea", "Oat", sep=" - ") → Output:
Bean - Pea - Oat
```

2.3.6.2.5 Print with End Parameter

By default, the print function automatically advances to the next line. However, the use of the optional *end* argument keeps the print function moving to the next line. The difference can be seen in the following examples.

```
print("The harvesting of Corn")
print("done.")
```

The output of the above print statements is shown below.

The harvesting of Corn
done.

```
print("The harvesting of Corn", end=" ... ")
print("done.")
```

The output of the above two print statements is ***The harvesting of Corn...done***. This example shows that the default value of the *end* argument is a new line character.

2.3.6.2.6 Print Mathematical Expressions

The print function can be used to display the results of mathematical expression(s) straightforwardly written in it. Examples are given below.

```
print(10 + 20) → Output: 30
```

```
crop_yield_in_field_X = 100
crop_yield_in_field_Y = 200
print(crop_yield_in_field_X + crop_yield_in_field_Y)
```

The output of the above print statement is ***300***.

2.3.6.2.7 Printing Multiple Items in a Single *Print* Statement

To print multiple items at once, separate them by comma(s) in the print statement and Python automatically inserts spaces between these items.

```
print("The agriculture yield of field X is", 200, "kg per acre")
```

The output of the above print statement is ***The agriculture yield of field X is 200 kg per acre***.

It is also important to mention here that the + and * characters in Python can also be used on strings for concatenation and repetition. The use of these operators with strings has been shown below.

```
# Use of * to print assignment symbol (=) 22 times
print("=" * 22)
crop_name = "Rice"

# Use of + to concatenate two strings
print("The crop name is "+ crop_name)

# Use of * to print assignment symbol (=) 22 times
print("=" * 22)
```

Below is the output of the above code snippet.

```
======================
The crop name is Rice
======================
```

2.3.6.2.8 Printing by Specifying Decimal Places

The *format* function of strings can be used to restrict certain place values after the decimal point.

```
price = 0.7515
print("Price per kg: ${:.2f}".format(price))
```

The output of the above print statement is ***Price per kg: $0.75***.

2.4 Programming Errors and Error Handling

2.4.1 Types of Programming Errors

In general, programming errors can be classified into three categories, i.e., *Syntax Errors*, *Runtime Errors*, and *Logical Errors*.

Syntax Errors are the errors generated by violating the rules of Python coding and are usually easy to detect. Incorrect indentation, mistyped, misplaced, or missing Python keywords, mismatched or omitted punctuation (i.e., parentheses, brackets, braces, quotes, commas, colons, etc.), use of Illegal characters in variable names, and incorrect use of the assignment (=) operator are common examples of syntax errors in Python.

Runtime Errors (also called *Exceptions*) occur during program execution. In simple words, these errors occur when the Python interpreter is executing a code that is syntactically correct but encounters an unexpected condition that causes the program to terminate abruptly.

Logical Errors occur when code executes successfully (without any Syntax or Runtime errors) but produces incorrect output. These types of errors are difficult to identify in a program because these are due to mistakes in the program's logic.

To understand the existence of Syntax, Logical, and Runtime errors in Python programs, consider the following example with its erroneous implementation in Listing 2.6 and correct implementation in Listing 2.7 (after the removal of errors in Listing 2.6).

Example: Write a Python program to calculate the total cost of fertilizing an agricultural field. The program should take the fertilizer cost per kilogram, the total amount of fertilizer in kilograms to fertilize one hectare, and the total area of the agricultural field (in hectares) from the user. The formula for calculating the total cost of fertilizer application to the agricultural field is

$$\text{Total Cost} = \text{Cost per kg} \times \text{kg per Hectare} \times \text{Field Area}\left(\text{Hectares}\right).$$

The Python solution (with errors) for this problem is in Listing 2.6.

Listing 2.6	Program (with errors) to calculate the total cost of fertilizing an agricultural field

```
'''
=============================================================
Python program to calculate the total cost of fertilizing a field
=============================================================
'''
# Input the cost of fertilizer per kilogram
cost_per_kg = float(inut("Enter the cost of fertilizer: ")
# Input the number of kilograms required per hectare
kg_per_hectare = input("Enter kilograms per hectare:")
# Calculate the fertilizer cost per hectare
fertilizer_cost_per_hectare = cost_per_kg * kg_per_hectare
# Input the total area of the field in hectares
field_area = float(input("Enter the total field area: "))
# Calculate the total cost
total_cost = fertilizer_cost_per_hectare + field_area
# Display the total cost
print("Total cost fertilizing the field is:", total_cost)
```

When you execute the code mentioned in Listing 2.6, then you will see the following error message:

```
line 7
cost_per_kg = float(inut("Enter the cost of fertilizer: ")
         ^
SyntaxError: '(' was never closed
```

This is an example of a Syntax Error, and it can be fixed by just putting the closing parenthesis *")"* at the end of line 7. After fixing this error, when you execute this code again, you will see the following error message:

```
line 7
cost_per_kg = float(inut("Enter the cost of fertilizer: "))
                    ^^^^
NameError: name 'inut' is not defined. Did you mean: 'input'?
```

This is also a type of Syntax Error, and it can be fixed by correcting the spelling of "*input*" (at line 7). Once this error is fixed, running the code again and entering values for *cost_per_kg* and *kg_per_hectare* will result in the following error message:

```
line 11
fertilizer_cost_per_hectare = cost_per_kg * kg_per_hectare
                              ~~~~~~~~~~~~^~~~~~~~~~~~~~~~
TypeError: can't multiply sequence by non-int of type 'float'
```

This error is known as *TypeError*, which is one of the examples of Runtime Errors (or Exceptions). To deal with these types of messages, there is a special mechanism known as Exception Handling, which has been discussed separately in the next Sect. 2.4.2. However, to avoid this error message, line 9 can be modified by using the *float()* function as follows:

```
kg_per_hectare = float(input("Enter kilograms per hectare:"))
```

After correcting the code with the above-provided code line, executing the program will display the output after entering values for *cost_per_kg*, *kg_per_hectare*, and *field_area*. However, this output will be wrong because of a logical error at line 15 in Listing 2.6. The logical error is due to the incorrect implementation of the formula. The correct formula to calculate the total cost should be *total_cost = fertilizer_cost_per_hectare * field_area* but it has been implemented as *total_cost = fertilizer_cost_per_hectare + field_area*. This is an example of a logical error, and these types of errors are challenging to find as the compiler or interpreter does not provide any hint for these types of errors. This error can be corrected by replacing the "+" with "*" (as shown in Listing 2.7).

Listing 2.7 Program (without errors) to calculate the total cost of fertilizing an agricultural field

```
'''
==============================================================
Python program to calculate total cost of fertilizing a field
==============================================================
'''

# Input the cost of fertilizer per kilogram
cost_per_kg = float(input("Enter the cost of fertilizer: "))

# Input the number of kilograms required per hectare
kg_per_hectare = float(input("Enter kilograms per hectare:"))

# Calculate the fertilizer cost per hectare
fertilizer_cost_per_hectare = cost_per_kg * kg_per_hectare

# Input the total area of the field in hectares
field_area = float(input("Enter the total field area: "))

# Calculate the total cost
total_cost = fertilizer_cost_per_hectare * field_area

# Display the total cost
print("Total cost fertilizing the field is:", total_cost)
```

2.4.2 Exception Handling

Although the code in Listing 2.7 is modified by applying the *float()* function at line 9 (of Listing 2.6) to prevent this Runtime *TypeError;* however, relying solely on such fixes is not the most effective approach. To ensure the program does not terminate unexpectedly due to runtime errors, it is essential to implement a proper mechanism that is known as exception-handling mechanism. Proper runtime error or exception handling using the *try-except* block shown below not only prevents abrupt failures but also enhances the program's reliability and user experience.

```
try:
  # code that may cause exception
except:
  # code to run when exception occurs
```

From the above syntax, it becomes evident that it is good practice to place the code that might generate an exception inside the *try* block that is followed by an *except* block. When a runtime error or an exception occurs, it is caught by the *except* block. It is important to note that the *except* block cannot be used without the *try* block. Considering the basic examples of *ZeroDivisionError* and *ValueError*, the use of the *try... except* block has been explained in Listing 2.8.

Listing 2.8	Python program demonstrating the exception handling mechanism

```
'''
============================================================
Example Python program to handle 3 types of exceptions, i.e.,
- Divide By Zero Exception (raised when any number is
divided by 0)
- Value Error Exception (raised when the operation is performed
to an inappropriate value)
============================================================
'''

# Start of try-except block
try:
    # Ask the user to input values for two numbers
    num_1 = float(input("Enter the Number 1 Value: "))
    num_2 = float(input("Enter the Number 2 Value: "))

    # Division calculation to get results
    result = num_1 / num_2
    print("The result is:", result)

except ZeroDivisionError:
    print("Error: Division by zero is not allowed.")
except ValueError:
    print("Error: Please enter valid numeric values.")

print("Doing other necessary tasks…")
print("Finish")
```

To understand the output of this program, consider the following example scenarios.

Scenario 1: If the user enters two numeric values, i.e., 11 and 9, for num_1 and num_2, respectively, then the output will be

```
Enter the Number 1 Value: 11
Enter the Number 1 Value: 9
The result is: 1.2222222222222223
Doing other necessary tasks…
Finish
```

Scenario 2: If the user enters a non-numeric value for any variables (num_1 or num_2), then the *except* block associated with *ValueError* will execute. For example, if the user enters "9A" value (i.e., a non-numeric value) for the num_1 variable, then the output will be

```
Enter the Number 1 Value: 9A
Error: Please enter valid numeric values.
Doing other necessary tasks...
Finish
```

Scenario 3: If the user enters a 0 value for the num_2 variable, then the *except* block associated with *ZeroDivisionError* will be executed and the output will be

```
Enter the Number 1 Value: 5
Enter the num_2: 0
Error: Division by zero is not allowed.
Doing other necessary tasks...
Finish
```

It is important to note that in the absence of *try ... except* block in this example code,

(a) The output for Scenario 2 will be

```
num_1 = float(input("Enter the Number 1 Value: "))
        ^^^^^^^^^^^^^^^^^^^^^^^^^^^^^^^^^^^^^^^^^^
ValueError: could not convert string to float: '9A'
```

(b) The output for Scenario 3 will be

```
result = num_1 / num_2
         ~~~~~~^~~~~~
ZeroDivisionError: float division by zero
```

The outputs (a) and (b) above show that in both scenarios (without *try...except* block) the program terminates abnormally, and the last two statements of Listing 2.8 are not executed.

2.5 Control Flow Structures (Selection and Iteration)

Control structures are one of the basic components of programming languages that are used to determine the execution flow of computer programs' instructions. Like other programming languages, Python has four types of control flow that are used to determine the way a program executes its instructions. In simple words, these structures allow the programmer to control the order in which statements are executed. These control structures are *Sequential*, *Selection*, *Iteration*, and *Jump* Control Flow Structures.

2.5.1 Sequential Flow Control

The simplest and default control structure is Sequential Flow Control, and as the name suggests, it executes the program instructions in series or sequentially one after another in the order they are written in a computer program.

2.5.2 Selection Flow Control

The Selection Flow Control structure allows the program to select or decide between different paths of execution based on some conditions. That is why these are also known as *Decision* or *Conditional* structures. In Python, these conditional structures are of four types, i.e.,

- *if statement* is used to execute a set of statements (or code block) when *if* condition or expression is evaluated as true.
- *if-else statement* executes a corresponding set of statements (or code block) when *if* condition or expression evaluation results in true, and another set of statements when *if* condition or expression evaluation results in false.
- *if-elif-else statement* is used to execute different sets of statements (or code blocks) under multiple conditions that evaluate in a sequence.
- *Ternary statement* is a concise way to implement a simple if-else statement in a single (code) line.

2.5.3 Iteration Control Flow

Iteration Control Flow involves the tasks of repetitive operations and allows the execution of a statement or a set of statements (or a block of code) multiple times, either a fixed number of times or until a condition is met. That is why it is also known as *Repetition* or *Loop* Control Flow Structures. The *while* loop and *for* loop are two primary Iterative constructs in Python. The *while* loop is ideal in conditions where the number of iterations or repetition of a block of code is unknown and depends on dynamic conditions, i.e., external factors in the form of user input or real-time data. The *for* loop is ideal for scenarios when the number of iterations is known beforehand.

2.5.4 *Jump Flow Control*

Jump Flow Control statements (i.e., *break* and *continue*) alter the normal flow of execution by jumping to another part of the computer program. For example, these statements can be used to terminate the current loop iteration or skip the current loop iteration.

The details of these control flow structures are available in Chap. 3.

2.6 Python Functions, Methods, and Modules

A function or method in a programming language is a block of code (or set of statements) that is designed to perform a particular task. Functions/methods are one of the most important building blocks of any programming language and allow developers to divide complex programs into smaller parts that are more organized and easier to handle. Python supports two types of functions or methods, i.e., *Built-In Functions/Methods* and *Custom Functions/Methods*. Built-In Functions/Methods are predefined functions/methods that come with a programming language to facilitate developers to perform the most common operations required in a computer program, i.e., input/output operations, operations on strings, mathematical operations, and so on. On the other hand, Custom Functions/Methods (also known as *User-Defined Functions/Methods*) allow developers to implement reusable code blocks to cater to specific application requirements. (Note: In programming languages, the terms Functions and Methods are used interchangeably but in Python there exist some subtle differences between these two that are discussed in Chap. 4). In Python, modules are Python files that contain Python statements in the form of functions, methods, classes, and so on that are imported and reused in other programs. Similar to Python functions/methods, modules in Python are of two types, i.e., *Built-In Modules* and *Custom Modules*. Built-In Modules are pre-written modules that come with the Python installation and are designed to handle common programming tasks to support the use of common operations in a computer program such as date/time operations, mathematical computations, string manipulation, file manipulation, and so on. On the other hand, Custom Modules (or *User-Defined Modules*) are created by developers to meet the specific needs of an application. Relevant details about Built-In functions, methods, and modules for string manipulation, mathematical calculations, along with the creation and usage of Custom functions, methods, and modules are available in Chap. 4.

2.7 Python Data Structures

In programming, data structure refers to a way of organizing, storing, and manipulating data in computer systems and plays an important role for developers to manage and access data efficiently and effectively. By combining primitive data types, these data structures simplify the higher-level data operations and complex data management tasks, i.e., searching, inserting, sorting, etc. Python's built-in data structures (including *Lists*, *Tuples*, Sets, and *Dictionaries*) not only enable developers to manage and manipulate data effectively but also help them to write clean and maintainable code. Each of these data structures is designed to fulfill a specific purpose. Details of these data structures have been discussed in Chap. 5.

2.8 File Handling

File handling is another important aspect of programming languages and it enables programmers to create, write, read, and manipulate files on the computer. Similar to other high-level languages, Python has a set of simplified yet powerful built-in functions to perform file-based operations on various types of binary and text files, i.e., images, audio, Word, Excel files, and so on that have been discussed in detail in Chap. 6.

2.9 Python Packages and Libraries

Like Python Modules, Python Packages and Python Libraries are used for organizing, managing, and utilizing code efficiently. The difference among these three is that a Python Module is a single file containing reusable Python code (mostly in the form of functions or classes), a Python Package is a directory or collection of related Python Modules, and Python Libraries are collections of related Python Packages and Python Modules. Built-in and custom packages and libraries not only save time and effort in development but also promote collaboration among developers. Important Python packages (i.e., *NumPy* and *Pandas*) and Python Libraries (i.e., *Matplotlib*, *Scikit-learn*) have been discussed in detail in Chaps. 7 and 8, respectively.

2.10 Case Studies

This section presents case studies along with their solution as Python programs to effectively demonstrate the application of Python's elementary components in different agricultural contexts.

Case Study 2.1: A farmer has a 5-hectare field of maize crop that produced a yield of 1300 kg of maize per hectare. This year, the market price for maize is US$ 0.60 per kg. For harvesting, the farmer needs to distribute 10 workers evenly across the three regions (North, West, East) of the same field, with leftover workers going to additional tasks. Moreover, the farmer also wants to know the budget for pesticide applications over time, with a yearly 5% increase rate in the cost of pesticide. Write a Python program to calculate

- Total revenue from crop yield
- Number of extra workers for additional tasks after the total workforce distribution
- Total increase in pesticide cost in the next 3 years, assuming the cost of pesticide application this year was US$ 100.

Solution:

The mathematical expressions to calculate **Total Revenue**, **Number of Extra Workers**, and **Future Pesticide Cost** are given as follows:

$$\text{Total Revenue} = \text{Field Size} \times \text{Crop Yield per Hectare} \times \text{Crop Market Price per kg}$$

$$\text{Extra Workers} = \text{Number of Workers}\,\textit{modulus}\,\text{Number of Field Regions}$$

$$\text{Future Pesticide Cost} = \text{Current Pesticide Cost} \times (1.05)^3$$

The Python program for this case study is given in Listing 2.9.

Listing 2.9	Program to calculate total revenue, extra workers, and future pesticide cost (Case Study 2.1)

```
'''
==============================================================
Program to calculate Total Revenue, Number of Extra Workers,
and Future Pesticide Cost
==============================================================
'''

# Given data
field_size = 5 # hectares
yield_per_hectare = 1300 # kg
market_price = 0.60 # $ per kg
available_workers = 10
field_regions = 3
initial_pesticide_cost = 100 # $
```

```
# Implementation of Mathematical given Expressions

# Calculating Crop Revenue (use of multiplication operator)
total_revenue = field_size * yield_per_hectare * market_price

# Calculating Number of Workers (use of modulus operator)
extra_workers = available_workers % field_regions

# Calculating Future Cost for Pesticide Spray
# (use of expontiation operator)
future_pesticide_cost = initial_pesticide_cost * (1.05 ** 3)

# Results' Output
print("Total Revenue of Maize Crop is: $", total_revenue)
print("Extra Workers Available for Additional Tasks:",
                                                    extra_workers)
print("Future Pesticide Cost in 3 Years: $",
                                   round(future_pesticide_cost, 2))
```

Case Study 2.2: A plant pathologist is interested in analyzing the spread of bacterial diseases in an apple orchard. The orchard has 900 trees, and the pathologist is monitoring the percentage of infected trees over a period of time. He observed that initially (in the first week) 3% of the trees are infected. However, afterward, each week, the infection spreads at a constant rate of 5% of the total trees (not cumulative). In light of the plant pathologist's observation, calculate

- number of infected trees at the beginning (in first week)
- total number of infected trees after 4 weeks.

Listing 2.10	Program to calculate total number of infected trees (Case Study 2.2)

```
# Given data
total_trees = 900
initial_infection_rate = 0.03 # 3 percent
weekly_rate = 0.05 # 5 percent
weeks = 4

# Calculating the initial number of infected trees
initial_infected = (total_trees * initial_infection_rate)

# Calculating the total number of infected trees
new_infected = (total_trees * (weekly_rate * (weeks - 1))
total_infected = initial_infected + new_infected

# Output results
print("Infected trees after week 1", initial_infected)
print("Infected trees after", weeks, "weeks:", total_infected)
```

Case Study 2.3: A horticulturist wants to calculate the number of orange fruit plants in a square system of planting within a given area. Write a Python program that takes user inputs for the total area (in square meters), row-to-row spacing, and the plant-to-plant spacing in meters. Using these inputs, the program should calculate and display the total number of plants that can be accommodated in the given area.

Listing 2.11	Program to calculate total number of plants in an orange orchard (Case Study 2.3)

```
'''
=================================================================
Program to calculate the number of orange plants in a given area.
=================================================================
'''
# Taking user input for area and plant-to-plant spacing
area_in_hectares = float(input("Enter area in Hectares: "))
row_spacing = float(input("Enter row spacing in meters: "))
plant_spacing = float(input("Enter plant spacing in meters: "))

# convert hectares in to square meters
area_in_sqm = area_in_hectares * 10000

# Calculating number of plants
# Use of floor division // to round the result
num_plants = area_in_sqm // (row_spacing * plant_spacing)

# Displaying result
print(f"Total orange plants to be planted: {num_plants}")
```

Case Study 2.4: To reduce the use of pesticides, an entomologist wants to determine the number of pheromone traps required for monitoring insect populations in an agricultural field. The traps are required to be placed in a uniform grid pattern with the same row/column specified spacing. Write a Python program that takes the field area and trap spacing as user inputs.

Listing 2.12	Program to calculate total pheromone traps in an agricultural field (Case Study 2.4)

```
'''
=================================================================
Program to calculate the total number of pheromone traps for the
monitoring of insect populations in an agricultural field
=================================================================
'''
```

```
# Taking user inputs for field area and trap spacing
field_area = int(input("Enter field area in square meters: "))
trap_spacing = int(input("Enter row/col space (in meters):"))

# Calculating the number of traps required
traps_required = field_area / (trap_spacing ** 2)
print(f"Number of pheromone traps needed: {int(traps_required)}")
```

In Listing 2.12, the exponentiation operator (**) is used because the trap spacing is the same in both row and column directions, and the use of *int* outside the *(traps_required)* is used to take the integer part of the calculated value.

2.11 Exercises

Problem 2.1 A horticulturist is planning to grow Tomatoes on a fixed 20-acre farm. However, before planting Tomatoes, the horticulturist wants to evaluate the total cost of cultivation (including the cost of seeds, fertilizers, and labor) and the net profit from the harvest. Write a Python program that should calculate the total cost of cultivation and net profit based on the following inputs provided by the horticulturist:

- Seed cost per acre,
- Fertilizer cost per acre,
- Labor cost per acre,
- Expected yield per acre (in kg), and
- Expected market price per kg.

The formulae for these calculations are given as follows.

$$\text{Total cost} = \text{Seed Cost} + \text{Fertilizer Cost} + \text{Labor Cost}(\text{per acre}) \times \text{Total Field Area}$$

$$\text{Total expected yield} = \text{Expected yield per Acre} \times \text{Total Field Area}$$

$$\text{Total revenue} = \text{Total expected yield} \times \text{Selling Price per kg}$$

$$\text{Net profit} = \text{Total Revenue} - \text{Total Cost}$$

Problem 2.2 To minimize the harm to the environment, it is important to control the use of pesticides in agricultural fields. Therefore, an entomologist wants to implement an Integrated Pest Management (IPM) strategy that involves the use of traps and biological control agents to control aphids on a fixed 10-acre vegetable farm that will ultimately improve the increase in yield per acre. However, before implementing the IPM strategy, he wants to know the total cost and expected increase in profit from increased yield per acre. Write a Python program (while considering the exception-handling mechanism to manage invalid inputs) to calculate the total cost of the selected IPM strategy along with the net profit by allowing the entomologist to use the following inputs:

- Cost of Monitoring Traps (CMT) in $
- Cost of Biological Control (CBC) in $
- Labor Cost per Acre (LCpA) in $
- Expected Selling Price per kg (ESP) in $
- Expected Increase in Yield (EIY) per acre in kg

The formulae for these calculations are given as follows:

$$\text{Total IPM Cost} = \text{CMT} + \text{CBC} + \text{LCpA} \times \text{Total Field Area}$$

$$\text{Total Increased Yield} = \text{EIY} \times \text{Total Field Area}$$

$$\text{Additional Revenue} = \text{Total Increased Yield} \times \text{ESP}$$

$$\text{Overall Profit} = \text{Additional Revenue} - \text{Total IPM Cost}$$

Problem 2.3 An agricultural extension officer is responsible for the distribution of a limited quantity of fertilizer among farmers in a specific region. To ensure fairness, the fertilizer must be distributed equally among the farmers, with any remaining quantity reserved for future use. Furthermore, the officer needs to calculate the nutrient concentration over time to guide farmers on sustainable fertilizer practices, using a specified degradation rate. Considering the exception-handling mechanism to manage invalid inputs, write a Python program that allows the officer to input the total fertilizer quantity (in kg), the number of farmers, and the degradation rate (suppose halve every year). The program should calculate the equal distribution of fertilizer per farmer, the leftover fertilizer for future use, and the nutrient concentration over a specified period.

Chapter 3
Control Flow Structures/Statements

3.1 Introduction

Computer programs do not always follow a simple, linear sequence of steps during execution. Often, programs are structured to make decisions, allowing multiple pathways based on changes in variables' values. Like most programming languages, Python control structures play a critical role in guiding the flow of a program and allow developers to execute code conditionally or repeatedly. Therefore, Python provides built-in control structures that support this type of flexible flow. These control structures mainly include *Conditional* and *Repetition* statements. Conditional (also known as *Selection* or *Decision*) statements (such as *if*, *if-else*, *if-elif-else*, etc.) enable the program to take different actions based on certain conditions. Repetition (also known as *Iteration* or *Loop*) statements (such as *for* and *while*) allow repetitive tasks until a specified condition is met. In addition, *Jump* flow control statements such as *break*, *continue*, *pass*, and *return* add flexibility within loops and decision structures, making code more efficient and dynamic for a wide range of applications. The Selection and Iterative control structures use *Boolean expressions* consisting of *Comparison* and *Logical* operators. Hence, to effectively use Selection or Iterative control structures, it is essential to first develop a solid understanding of Comparison operators, Logical operators, and Boolean expressions. This foundational knowledge assists you in constructing and evaluating conditions, which are key to guiding program flow with Conditional and Iterative statements.

M. A. Iqbal, *Python for Agriculturists*,
https://doi.org/10.1007/978-3-032-01442-9_3

3.2 Comparison Operators and Logical Operators

The six Comparison operators in Python are == *(equal to)*, != *(not equal to)*, > *(greater than)*, < *(less than)*, >= *(greater than or equal to)*, and <= *(less than or equal to)*. The three Logical operators are *and*, *or*, and *not*. The Comparison and Logical operators with their semantics have been shown in Table 3.1. The precedence chart of Python's operators is shown in Table 3.2.

*Note: If operators with the same precedence appear next to each other, their associativity determines the order in which they are evaluated. In Python, all binary operators (e.g., +, −, *, /, %) are left-associative, meaning they are evaluated from left to right. Moreover, parenthesis ( ), braces { }, and brackets [ ] take precedence over everything else.*

The code in Listing 3.1 demonstrates operator precedence in Python.

Table 3.1 Python comparison and logical operators

Operator type	Operator	Mathematical notation	Meaning	Example
Comparison operators	>	>	Greater than	Determine if rainfall exceeds the required level: Rainfall > 50
	>=	≥	Greater than or equal to	Check if a fertilizer's cost meets a threshold: Cost >= 100
	<	<	Less than	Identify fields with less than expected crop yield: Yield < 500
	<=	≤	Less than or equal to	Verify if the pest density is manageable: Pests <= 10
	==	=	Equal to	Check if two yields are the same: Yield_A == Yield_B
	!=	≠	Not equal to	Verify if costs for two pesticides differ: Cost_A != Cost_B
Logical operators	and	∧	Logical conjunction (returns True if both conditions are true)	Check if the yield is high and the cost is low: Yield > 500 and Cost < 100
	or	∨	Logical disjunction (returns True if at least one condition is true)	Determine if pests or diseases exceed limits: Pests > 20 or Diseases > 10
	not	¬	Logical not (reverses the Boolean value)	Verify if rainfall is not below the required level: not (Rainfall < 50)

Table 3.2 Operators' precedence in Python

Operator	Precedence
(Unary Plus and Minus) +, −	Highest
(Exponentiation) **	
Not (logical operator)	
(Multiplication, division, integer division, and remainder) *, /, //, %	↓
(Binary addition and subtraction) +, −	
(Comparison) <, <=, >, >=	
(Equality Comparison) ==, !=	
and (logical operator)	
or (logical operator)	
(Assignment and Compound Assignment operators) =, +=, -=, *=, /=, //=, %=	Lowest

Listing 3.1	Program demonstrating Operator Precedence in Python

```
'''
Python program for entomologists who are interested in
determining the correct dose of pesticide per hectare while
considering the factors of pest severity, temperature, and
rainfall.

Consider the following formula to calculate the correct
pesticide dose.
Required Pesticide Dose (L/ha) =
            (Pest Severity × X) - (Temperature/Y + Rainfall × Z)
where:
- X, Y, Z are crop-specific constants (assumed values of X, Y, and
  Z are 2, 5, and 0.3, respectively, in this example)
- Pest Severity is measured on a scale (e.g., 1 to 10)
- Temperature in °C
- Rainfall in mm
'''

# Taking inputs from the user
pest_severity = int(input("Enter pest severity level in %age: "))
temp = int(input("Enter temperature in Celsius: "))
rainfall = int(input("Enter rainfall in mm: "))

# Applying the formula to calcualte required pesticide dose
req_dose = pest_severity * 2 - (temp / 5 + rainfall * 0.3)

# Display result for required pesticide code
print(f"Required Dose: {req_dose:.2f}")
```

Output of Listing 3.1:

The self-explanatory output of Listing 3.1 is shown below.

Enter pest severity level in %age: 5
Enter temperature in Celsius: 20
Enter rainfall in mm: 10
Required Dose: 3.00

Explanation:
Considering the input values taken from the user, below is the explanation of how this expression is evaluated in Python

required_dose = pest_severity * 2 - (temperature / 5 + rainfall * 0.3) → Inside parenthesis first
required_dose = 5 * 2 - (**20/5** + 10 * 0.3) → Division due to left-associativity
required_dose = 5 * 2 - (4 + **10 * 0.3**) → Multiplication precedence is higher than addition
required_dose = 5 * 2 - (**4 + 3**) → Parenthesis precedence is higher
required_dose = **5 * 2** – 7 → Multiplication precedence is higher than subtraction
required_dose = **10 – 7** → Subtraction
required_dose = 3 → Result

3.3 Boolean Expressions

Comparison and Logical operators are used to evaluate relationships between values, resulting in Boolean or Conditional expressions that return a Boolean Value (either *True* or *False*). These Boolean or Conditional expressions form the basis of Conditional statements and Iterative control structures and allow computer programs to make decisions by determining whether certain conditions are met. Below are a few example code snippets (along with their self-explanatory outputs) related to the use of Boolean statements in Python.

Code Snippet 1: Use of basic comparison operators

```
fieldA_yield = 10
fieldB_yield = 20
print(fieldA_yield < fieldB_yield ) # → Output: True
print(fieldB_yield < fieldA_yield ) # → Output: False
print(fieldB_yield == fieldA_yield ) # → Output: False
print(fieldB_yield != fieldA_yield ) # → Output: True
```

Code Snippet 2: Use of ***and*** operator

```
soil_temp = 30
soil_humidity = 70
print(soil_temp > 25 and soil_humidity > 65) # → True
print(soil_temp < 25 and soil_humidity > 65) # → False
print(soil_temp > 25 and soil_humidity < 65) # → False
print(soil_temp < 25 and soil_humidity < 65) # → False
```

Code Snippet 3: Use of ***or*** operator

```
crop_type = "wheat"
season = "summer"
print(crop_type == "corn" or season == "summer")# → True
print(crop_type == "wheat" or season == "winter")# → True
print(crop_type == "wheat" or season == "summer")# → True
print(crop_type == "corn" or season == "summer")# → False
```

Code Snippet 4: Checking equality (==)

```
fieldA_soil_type = "clay"
fieldB_soil_type = "sandy"
print(fieldA_soil_type == "sandy") # → Output: False
print(fieldB_soil_type == "sandy") # → Output: True
```

Code Snippet 5: Use of ***not*** operator

```
is_raining = False
print(not is_raining) # True
```

These examples (code snippets) illustrate the use of simple Boolean expressions in Python statements. These Boolean expressions are used in Conditional and Iterative control structures to control the flow (or direction) of a Python program.

3.4 Conditional (Selection or Decision) Control Structures

Python's Conditional control structures enable a Python program to execute certain code blocks only when specific conditions are met. In other words, the Conditional control structures support decision-making within the code. There are five Conditional control flow structures in Python, i.e., *if*, *if-else*, *nested if-else*, *if-elif-else*, and *ternary* statements. Evaluating the Boolean expression(s) in these conditional control structures, Python can execute corresponding blocks of code based on true or false results. This allows the code to adapt its behavior dynamically. Additionally, nested and complex conditions enable more detailed branching, making conditional control structures essential for building adaptable programs that can handle a variety of scenarios. The use of all five conditional statements of Python (i.e., *if*, *if-else*, *nested if-else*, *if-elif-else*, and *Ternary*) has been discussed below.

3.4.1 if *Conditional Control Structure*

In Python, an *if* statement is used to execute code only if a specified condition evaluates to true. It forms the foundation of decision-making in programming by allowing code to respond to certain conditions. If the condition is met, the code block under the *if* statement is executed; if not, the program skips that block entirely. The general syntax of an *if* statement is

```
if boolean_expression:
    statement(s) # body of if statement
```

Here, the *Boolean expression* (containing Comparison and/or Logical operators) is used to evaluate true or false values, and the body is executed only if the evaluation result is true. In case of multiple statements, the *if* body statements must be grouped with one level of indentation. Listings 3.2 and 3.3 demonstrate the use of *if* conditional control.

Listing 3.2	Program demonstrating the use of ***if*** conditional structure with Comparison operator

```
'''
Python program for farmers who are interested in checking if the
current soil moisture level is adequate or not to skip watering.
'''

soil_moisture_threshold = 50

soil_moisture = eval(input("Enter Soil Moisture (%age): "))

if soil_moisture > soil_moisture_threshold:
    print("Soil is sufficiently moist, no need to water.")

print("End")
```

Output of Listing 3.2:

(a) If the user enters current moisture level value greater than 50, the output will be
Soil is sufficiently moist, no need to water.
End

(b) When the user enters current moisture value as 50 or less than 50, the output will be
End

Listing 3.3	Example demonstrating the use of *if* conditional structure with Comparison and Logical operators

```
'''
Python program for an agronomist who is interested in checking if
both the current soil moisture level and current temperature are
ideal for crop sowing.
'''

soil_moisture = eval(input("Enter Soil Moisture (%age): "))
temperature = eval(input("Enter Temperature (in °C): "))

if soil_moisture >= 50 and temperature >= 18:
    print("Conditions are ideal for sowing.")

print("End")
```

Output of Listing 3.3:

(a) If the user enters the current moisture level value as 50 or more than 50 AND current temperature as 18 or more than 18, then the output will be
Conditions are ideal for planting.
End

(b) If the user enters the current moisture value as less than 50 AND current temperature as 18 or more than 18, the output will be
End

(c) If the user enters the current moisture value as 50 or more than 50 AND current temperature as less than 18, then the output will be
End

(d) If the user enters the current moisture value as less than 50 AND current temperature as less than 18, then the output will be
End

3.4.2 if-else *Conditional Control Structure*

The *if-else* statement adds an alternative action to the simple *if* control structure. If the condition or Boolean expression results in true, the program executes the *if* block; if false, it executes the *else* block. This structure is useful for handling two possible outcomes, such as processing user input based on specific requirements. The general syntax of an *if-else* statement is

```
if boolean_expression
    # Code to execute if boolean_expression is True
    Statements(s1)
else:
    # Code to execute if boolean_expression is False
    Statements(S2)
```

Listing 3.4 demonstrates the use of *if-else* conditional control.

Listing 3.4	Program demonstrating the use of *if-else* conditional structure

```
'''
Python program for soil scientists who are interested in deciding
whether Potassium fertilizer is needed to fertilize a certain
crop or not.
'''

# Enter Potassium level in soil(unit mg/L)
k_nutrient_level = eval(input("Enter Potassium Level in Soil: "))

if k_nutrient_level <= 100:
    print("Potassium level is low, apply fertilizer.")
else:
    print("Potassium level sufficient, no fertilizer needed.")
```

Output of Listing 3.4:

(a) If the user enters the current Potassium level as 100 or less than 100, then the output will be
Potassium level is low, apply fertilizer.

(b) If the user enters the current Potassium level above100, then the output will be
Potassium level sufficient, no fertilizer needed.

3.4.3 Nested if-else *Conditional Control Structure*

The *nested if-else* structure permits the existence of an *if* statement within another *if* or *else* statement. In this way, multiple layers of conditional logic can be written within a program. The *nested if-else* structure is helpful when decisions depend on multiple conditions, and in such cases, each conditional logic layer generates a more specific result. Listing 3.5 demonstrates the use of *nested if-else* conditional control.

Listing 3.5 | Program demonstrating the use of *nested if-else* conditional structure

```
'''
Python program for greenhouse managers to check temperature and
humidity levels to take decisions about greenhouse ventilation.
'''

temp = eval(input("Enter Greenhouse Temp.(°C): "))
humidity = eval(input("Enter Greenhouse Humidity (%age)"))

if temp >= 30:
    if humidity >= 70:
        print("Turn on fan and open ventilation windows.")
    else:
        print("Only turn on fan for ventilation.")
else:
    print("Ventilation is not necessary.")
```

Output of Listing 3.5:

(a) If the user enters the current temperature level as less 30, then the output will be
Ventilation is not necessary.

(b) If the user enters the current temperature level as 30 or greater than 30, then the control will move to the nested if statement to check whether the humidity level is greater than or equal to 70 or not. If the entered humidity level is greater than or equal to 70, then the output will be
Turn on fan and open ventilation windows.

(c) On the other hand, if the entered humidity level is less than 70, then the output will be
Only turn on fan for ventilation.

3.4.4 if-elif-else *Conditional Control Structure*

The *if-elif-else* conditional control structure is beneficial when there are multiple conditions to evaluate in a sequence. If one condition or Boolean expression results in true, then the corresponding code block (or set of statements) is executed, and the rest are ignored. This control structure is particularly useful when dealing with various specific conditions rather than a simple binary choice. The general syntax of the *if-elif-else* structure is

```
if boolean_expression1
# S1 statements execute if boolean_expression1 is True
    Statements({S1})
elif boolean_expression2:
# S2 statements execute if boolean_expression2 is True
    Statements({S2})
...
elif boolean_expressionN
# SN statements execute if boolean_expressionN is True
    Statements({SN})
else:
# This block executes if above expressions are False
```

Listings 3.6 and 3.7 demonstrate the use of ***if-elif-else*** conditional control.

Listing 3.6	Program demonstrating the use of ***if-elif-else*** conditional structure

```
'''
Python program that evaluates multiple conditions (considering
different combinations of current soil moisture level and
possibility of rain in the near future) that are required for the
proper working of a smart irrigation system.
'''

soil_moisture = eval(input("Enter soil moisture (%age): "))
rain_forecast = True

if soil_moisture < 30 and not rain_forecast:
    print("Irrigate the crops.")
elif soil_moisture < 30 and rain_forecast:
    print("Wait for rainfall.")
elif soil_moisture == 30 and not rain_forecast:
    print("Monitor soil moisture continuously.")
else:
    print("No action needed.")
```

Output of Listing 3.6:

(a) If the entered (representing current) moisture level is less than 30 and rain is not expected, then the statement under the first if condition will be executed, and the output will be
Irrigate the crops.

(b) If the entered (representing current) moisture level is less than 30 and rain has been forecasted, then the statement under the first elif condition will be executed, and the output will be
Wait for rainfall.

(c) If the entered (representing current) moisture level is 30 and rain is not in the forecast, then the statement under the second elif condition will be executed, and the output will be
Monitor soil moisture continuously.

(d) If the entered (representing current) moisture level is greater than 30 (irrespective of the fact that the rain is in the forecast or not), then the statement under the else condition will be executed, and the output will be
No action needed.

Listing 3.7	Program demonstrating the use of ***if-elif-else*** conditional structure

```
'''
Python program to display caring advice for plants in the garden
while considering their growth stage levels, i.e., (1) Seedling,
(2) Vegetative, and (3) Flowering
'''
growth_stage = eval(input("Enter Plant Growth Stage (1-3):"))

if growth_stage == 1:
    print("Water lightly & protect from sunlight.")
elif growth_stage == 2:
    print("Ensure adequate watering & sunlight.")
elif growth_stage == 3:
    print("Ensure adequate watering & nutrient supply.")
else:
    print("Monitor crop health regularly.")
```

Output of Listing 3.7:

Depending on the entered current growth level value, the respective statement will be executed. For example, if the user inputs 2 for the growth level, then the output will be

Ensure adequate watering & sunlight.

3.4.5 *Comparison of* if-elif-else *vs* nested if-else

Consider the implementations (Listings 3.8 and 3.9) of the following case study to better understand the advantages of using the *if-elif-else* conditional control structure over using the *nested if-else* conditional control structure.

Case Study 3.1: An agricultural research officer (ARO) needs a program to check the up-to-date fertilizer requirements in an agricultural field based on the field's current soil nutrient level (low, medium, high) and crop growth stage (vegetative or flowering). The appropriate level of fertilizer application to the agricultural field can be determined based on the following criteria:

- If the nutrient level is low and the growth stage is vegetative, then a high amount of fertilizer is required
- If the nutrient level is low and the growth stage is flowing, then a medium amount of fertilizer is required
- If the nutrient level is medium and the growth stage is vegetative, then a medium amount of fertilizer is required
- If the nutrient level is medium and the growth stage is flowing, then a low amount of fertilizer is required
- If the nutrient level is high and the growth stage is vegetative, then a low amount of fertilizer is required
- If the nutrient level is high and the growth stage is flowing, then no fertilizer is required

Listing 3.8	Example demonstrating the use of *if-elif-else* conditional structure

```
'''
Python program implemented using if-elif-else construct to
determine the fertilizer requirement based on nutrient level and
crop growth stage
'''

soil_nutrient = input("Enter soil nutrient level: ")
growth_stage = input("Enter crop growth stage: ")

# Implementation using if-elif-else
if soil_nutrient == "low" and growth_stage == "vegetative":
    print("High fertilizer dose required.")
elif soil_nutrient == "low" and growth_stage == "flowering":
    print("Medium fertilizer dose required.")
elif soil_nutrient == "medium" and growth_stage == "vegetative":
    print("Medium fertilizer dose required.")
```

```
elif soil_nutrient == "medium" and growth_stage == "flowering":
    print("Low fertilizer dose required.")
elif soil_nutrient == "high" and growth_stage == "vegetative":
    print("Low fertilizer dose required.")
elif soil_nutrient == "high" and growth_stage == "flowering":
    print("No fertilizer required.")
else:
    print("Incorrect nutrient level or growth stage.")
```

Listing 3.9	Example demonstrating the use of ***nested if-else*** conditional structure

```
'''
Python program implemented using nested if-else construct to
determine the fertilizer requirement based on nutrient level and
crop growth stage
'''

soil_nutrient = input("Enter soil nutrient level: ")
growth_stage = input("Enter crop growth stage: ")

# Implementation using Nested if-else

if soil_nutrient == "low":
    if growth_stage == "vegetative":
        print("High fertilizer dose required.")
    else:
        if growth_stage == "flowering":
            print("Medium fertilizer dose required.")
        else:
            print("Entered growth stage is not correct.")
else:
    if soil_nutrient == "medium":
        if growth_stage == "vegetative":
            print("Apply medium fertilizer dosage.")
        else:
            if growth_stage == "flowering":
                print("Apply low fertilizer dosage.")
            else:
                print("Entered growth stage is not correct.")
    else:
        if soil_nutrient == "high":
            if growth_stage == "vegetative":
                print("Apply low fertilizer dosage.")
            else:
```

```
        if growth_stage == "flowering":
            print("No fertilizer required.")
        else:
            print("Entered growth stage is not correct.")
    else:
        print("Entered soil nutrient level is not correct.")
```

From Listings 3.8 and 3.9, it becomes evident that the implementation of *if-elif-else* should be preferred for the following reasons:

- The readability of *if-elif-else* is better as the flow of decisions is linear and it avoids deeply nested blocks. That is why it is easier to follow the logic using *if-elif-else*. In the case of *nested if-else* conditions, the logic or flow of decisions is harder to follow due to deeply nested conditions.
- Maintenance (debugging or locating errors in the program) is easier. Moreover, new options can be added without reconstructing the existing *if-elif-else* structure. On the other hand, in the case of *nested if-else* structures, complexity is high due to nested block structures. That is why it is harder to trace and ultimately makes debugging more challenging.
- The efficiency of the *if-elif-else* structure is better as it evaluates conditions sequentially. Evaluation stops as soon as a condition is matched and therefore it avoids unnecessary evaluations once a condition is met.

3.4.6 Ternary Conditional Control Structure

A ternary or conditional expression is a concise way to implement a simple *if-else* statement in a single (code) line. In Python, the syntax of the Ternary conditional control structure is

```
[value_if_true] if [boolean_expression] else [value_if_false]
```

Listings 3.10 and 3.11 illustrate the use of the Ternary or conditional structure.

Listing 3.10	Program demonstrating the use of ***Ternary*** or conditional structure

```
'''
Python program that helps agriculturists to make decision whether
it is safe to start planting or wait to plant on the basis of
given temperature value.
'''
```

```
temp = eval(input("Enter Temperature in Celcius: "))
advice = "Safe to plant" if temp > 25 else "Wait to plant"
print(advice)
```

Output of Listing 3.10:

(**a**) If the entered temperature is greater than 25, then the output will be
Safe to Plant

(**b**) If the entered temperature is 25 or less than 25, then the output will be
Wait to plant

Listing 3.11	Program demonstrating the use of ***Ternary*** conditional structure

```
'''
Python program that helps agriculturists to make decisions
whether it is safe to apply a pesticide or not on the basis of
the provided pest density per plant and current wind speed.
'''

# Pest density here refers the number of pests per plant
pest_density = eval(input("Enter Pest Density: "))

# wind speed unit is miles per hour (mph)
wind_speed = eval(input("Enter Wind Speed: "))

advice = "Start Spray" if pest_density >= 10 and wind_speed <=10
                                    else "Do not apply pesticide"
print(advice)
```

Output of Listing 3.11:

(**a**) If the entered pest density is 10 or greater than 10 and the entered wind speed is 10 or less than 10, then the output will be
Start Spray

(**b**) If the entered pest density is less than 10 and the entered wind speed is greater than 10, then the output will be
Do not apply pesticide

3.4.7 *Common Indentation Errors in Conditional Statements*

Most common errors in selection statements are caused by incorrect indentation. Using correct indentation is very important as it ensures the proper code execution without syntax and logical errors. The common indentation errors in Python conditional statements (along with the correct solution) have been illustrated in Listings 3.12–3.14.

Listing 3.12	Error type: missing indentation

Code with missing indentation error

```
rainfall = 30
if rainfall > 20:
print("Watering not required")  # Error: line not indented
```

Code with Correction

```
rainfall = 30
if rainfall > 20:
    print("Watering not required")
```

Listing 3.13	Error type: inconsistent indentation

Code with inconsistent indentation error

```
soil_moisture = 15
if soil_moisture < 10:
    print("Irrigation needed")
    else:                            # Inconsistent indentation
      print("Irrigation not needed") # Inconsistent indentation
```

Code with correction

```
soil_moisture = 15
if soil_moisture < 10:
    print("Irrigation needed")
else:
    print("Irrigation not needed")
```

Listing 3.14	Error type: incorrect indentation in *nested if-else*

Code with incorrect indentation error

```
soil_moisture = "low"
temp = "medium"
if soil_moisture == "low":
   if temp == "hot":
      print("Irrigation ON: High level")
   elif soil_moisture == "medium":
        if temp == "moderate" or temp == "low":
           print("Irrigation OFF")
        else:
           print("Irrigation ON: Low level")
        elif soil_moisture == "high": # Incorrect indentation
           print("Irrigation OFF: Soil moisture sufficient")
   else: # This else should belong to the first if (Logical Error)
        print("Invalid input")
```

Note: Incorrect indentation of nested if statements sometimes leads to unexpected results or logical errors, as is the case of the last else statement in this code snippet.

Code with correction

```
soil_moisture = "low"
temp = "medium"
if soil_moisture == "low":
        if temp == "hot":
            print("Irrigation ON: High level")
        elif soil_moisture == "medium":
             if temp == "moderate" or temp == "low":
                print("Irrigation OFF")
             else:
                print("Irrigation ON: Low level")
        elif soil_moisture == "high":
            print("Irrigation OFF: Soil moisture sufficient")
else:
  print("Invalid input")
```

From the above examples, it becomes evident that improper indentation in Python Control Flow structures disrupts the flow of execution and leads to syntax and logical errors. Moreover, improper indentation increases the debugging complexity of Python code.

3.5 Iterative Flow Control Structures

Many real-world problems demand a repetition mechanism where the same set of statements must be executed multiple times. For such requirements, programming languages including Python have *Iterative* (also known as *Repetition* or *Loops*) control structures. Python primarily offers two types of Iterative Flow Control Structures: *while* loop and *for* loop. These loops are one of the essential components of Python to execute a code block (or set of statements) multiple times. Moreover, combined with Jump control statements (i.e., *break* and *continue*), these Iterative control structures offer flexibility for effectively managing repetitive tasks.

3.5.1 *The* While *Loop*

The *while* loop is an Iterative Control Flow Structure that is condition-based and is suitable when the number of iterations is not known in advance. This loop continues to execute as long as the specified condition(s) remain true. The general syntax of *while* loop is

```
while boolean_espression:
    Statement(s) # loop body
```

Like conditional control structures, the *Boolean expression* in a *while* loop (using comparison and/or logical operators) is used to evaluate true or false values. The statement(s) in the *while* loop body execute(s) until the evaluation result of the Boolean expression remains true. In the case of multiple statements, the while loop body statements must be grouped with one level of indentation. Listing 3.15 demonstrates the execution flow of a *while* loop.

Listing 3.15	Example demonstrating the execution of ***while*** loop

```
'''
=================================================================
Python program describing the working of the while loop
considering the example of printing values in ascending and
descending order.
=================================================================
'''
```

```
1. # To print values in ascending order
2. counter1 = 1
3. print("Counter 1 Values in Ascending Order")
4. while counter1 <= 5:
5.     print("Count 1:", counter1, end = ",")
6.     counter1 += 1 # Increment count by 1
7. print("End of while-loop execution")
8.
9. # To print values in descending order
10. counter2 = 5
11. print("Counter 2 Values in Descending Order")
12. while counter2 > 0:
13.     print("Count 2:", counter2, end = ",")
14.     counter2 -= 1 # Decrement count by 1
15. print("End of while-loop execution")
```

Output of Listing 3.15:

Counter 1 Values in Ascending Order
Count 1: 1,Count 1: 2,Count 1: 3,Count 1: 4,Count 1: 5,End of while-loop execution
Counter 2 Values in Descending Order
Count 2: 5,Count 2: 4,Count 2: 3,Count 2: 2,Count 2: 1,End of while-loop execution

Explanation:
Step 1: The program execution starts from line 2, where the counter 1 variable is created and initialized with value 1. Ultimately, this variable is used as a while loop counter.
Step 2: After that, the control moves to the next line (line 3), and the statement "Counter 1 Values in Ascending Order" is displayed on the output.
Step 3: At line 4, the Boolean expression condition (counter 1 <= 5) is checked before each iteration, and if it evaluates to true, the statements in the loop body execute.
Step 4: At line 5, the first statement of the while loop body is executed and "Count 1: 1," is displayed, and here the end="," ensures that the next output will stay on the same line, separated by a comma.
Step 5: At line 6, the counter value is incremented by 1 after the first iteration.
Step 6: After the increment at line 6, control shifts to line 4 again to evaluate the Boolean expression of the while loop. If it evaluates to true, then all execution steps (from step 4–6) will be executed again, and the counter will be incremented by 1 again until the Boolean expression of the while loop evaluates to false. When the Boolean expression evaluates false, then the loop execution terminates, and the statement "End of while-loop execution" is printed on the output.

After printing this statement, the control moves to Line 10, where a new variable counter2 is created and initialized with the value 5. Next, the execution proceeds with the second while loop, following the same process as the first while loop. However, the only difference is that this time, the counter is decremented in each iteration instead of being incremented.

A *while* loop is also used to validate the user input, which assists the programmer in implementing logic related to the acceptance of only correct input from the user, as demonstrated in Listing 3.16.

Listing 3.16	Example demonstrating the execution of ***while*** loop for accepting only correct input from the user

```
'''
================================================================
Python program demonstrating the working of the while loop for
validating and accepting the correct input from the user.
================================================================
'''
1. check = True
2. while check == True:
3.     password = input("Enter Password to Access Record: ")
4.     if password == "P79qRX13":
5.         print("Access is Granted for Crop Yield Record!")
6.         check = False
7.     else:
8.         print("Incorrect Password, Access Denied!")
```

Sample Output of Listing 3.16:

Enter Password to Access Record: Wieeujd
Incorrect Password, Access Denied!
Enter Password to Access Record: P79qRX95
Incorrect Password, Access Denied!
Enter Password to Access Record: P79qRX13
Access is Granted for Crop Yield Record!

Explanation:

Step 1: The program execution starts from line 1, where the check variable is created and initialized with the Boolean value True. This variable is declared for an infinite *while* loop execution.

Step 2: At line 2, the Boolean condition is evaluated before each iteration. As it evaluates true here, the statements in the loop body will execute.

Step 3: At line 3, the user is asked to enter the password.

Step 4: At line 4, the entered password is compared with the given value "P79qRX13," and if this comparison evaluates to true, then the "Access is Granted for Crop Yield Record!" statement will be displayed and to stop *while* loop next iterations, variable *check* will be set to False. Otherwise, if the user enters any value other than the given value, then the "Incorrect Password, Access Denied!" statement will be printed, and control moves back to check the *while* loop condition for the next iteration.

One of the incarnations of the *while* loop implementation is the *nested while* loop. In programming languages, a nested loop refers to a loop inside another loop's body. Considering this definition of nested loops, a *nested while* loop in Python is a *while* loop in another *while* loop. From an execution point, for each iteration of the outer loop, all iterations of the inner loop are executed. Nested loops are useful when multilevel iterations are required to be performed, e.g., in handling multidimensional data. In an agricultural context, it helps agriculturists to implement group-related tasks, i.e., analyzing crop growth across different seasons or in different parts of an agricultural field, or for monitoring pest infestation per plant in different plots. Listing 3.17 demonstrates the use of a nested while loop in an agricultural context.

Listing 3.17	Example demonstrating the working of ***nested while*** loop to calculate total fertilizer and pesticide cost across multiple farms

```
'''
==================================================================
Python program demonstrating the working of the nested while loop
to calculate total fertilizer and pesticide cost across multiple
farms, where each farm consists of multiple fields.
==================================================================
'''
# Get total number of farms from user
total_farms = int(input("Enter the number of farms: "))

# Declare and initialize variables
total_farm_cost = 0
total_cost = 0
farm_counter = 1
fertilizer_cost_per_hectare = 10
pesticide_cost_per_hectare = 20

# Outer loop to calculate total cost at farm level
while farm_counter <= total_farms:
    print("Farm Number: ", farm_counter)

    # Get number of fields in the current farm
    num_fields = int(input(f"Total fields in Farm {farm_counter}:"))
    field_counter = 1

    # Inner loop to calculate total cost at field level
    while field_counter <= num_fields:
        print("Field number :", field_counter)
        field_area = float(input("Enter field area (hectares): "))
        fertilizer_cost = fertilizer_cost_per_hectare * field_area
        pesticide_cost = pesticide_cost_per_hectare * field_area

# Calculate total cost for the current field
```

```
        field_cost = fertilizer_cost + pesticide_cost

# Display total cost of one field
        print("Field ", field_counter, " Cost: $", field_cost)

        # Add field total to overall total cost
        total_farm_cost += field_cost

        field_counter += 1 # Increment to move next field
    farm_counter += 1 # Increment to move next farm

    # Display total cost on one farm
    print("Total farm cost:", total_farm_cost)
    total_cost += total_farm_cost

# Display total cost of all farms
print("Total cost for all farms is $ ", total_cost)
```

Sample Output of Listing 3.17:

Enter the number of farms: 2
Farm Number: 1
Total fields in Farm 1: 2
Field number: 1
Enter field area (hectares): 6
Field 1 Cost:$ 180.0
Field number: 2
Enter field area (hectares): 5
Field 2 Cost: $ 150.0
Total farm cost: $ 330.0
Farm Number: 2
Total fields in Farm 2: 1
Field number: 1
Enter field area (hectares): 7
Field 1 Cost: $ 210.0
Total farm cost: $ 540.0
Total cost for all farms is $ 870.0

Explanation:
The program first prompts the user to enter the total number of farms and then the required variables are declared and initialized. Next, the outer while loop iterates to calculate the cost of fertilizer application and pesticide spray for each farm. However, before calculating each farm's total cost, the inner while loop iterates to get each field size from the user and calculates the cost of fertilizer application and pesticide spray for all fields on the farm. At the end, after calculating the fertilizer and pesticide cost for all farms, the program calculates and displays the total fertilizer and pesticide cost across all farms.

3.5.2 *The* for *Loop*

The *for* loop is an Iterative control structure that is used when you want to execute a block of code (set of statements) for a fixed number of times. This count-based loop is suitable for scenarios where users already know exactly how many times the loop should be executed. The three general syntaxes of Python ***for*** loop are as follows:

First Syntax

```
for variable_name in range(value):
    Statement(s)# for loop body
```

Second Syntax

```
for variable_name in range(initial_value, end_value):
    Statement(s)# for loop body
```

Third Syntax

```
for variable_name in range(initial_value, end_value, step_value):
    Statement(s)# for loop body
```

The first syntax represents a Python *for* loop that iterates over a sequence of numbers generated by the *range()* function. The *range(value)* function here in this statement generates numbers from 0 to *value* −1. Therefore, the statement(s) in the body of this type of *for* loop are executed *value* times starting from 0, as shown in Listing 3.18.

Listing 3.18	Generic example of *for* loop execution with *range(value)* function

```
for counter in range(6):
    print("Counter Value", counter)
```

Output of Listing 3.18:

```
Counter Value 0
Counter Value 1
Counter Value 2
Counter Value 3
Counter Value 4
Counter Value 5
```

The second syntax indicates that the *for* loop executes the statement(s) in its body for *end_value – initial_value* times (starting from *initial_value* to *end_value −1*, incrementing by 1 in each iteration), as shown in Listing 3.19.

Listing 3.19	Generic example of *for* loop execution with *range (initial_value, end_value)* function

```
for counter in range(1, 6):
    print("Counter Value", counter)
```

Output of Listing 3.19:

```
Counter Value 1
Counter Value 2
Counter Value 3
Counter Value 4
Counter Value 5
```

In the third syntax, the *range(initial_value, end_value, step_value)* function returns the sequence from *initial_value* to *end_value −1*, but each time with an increment or decrement as explicitly mentioned as *step_value*, as shown in Listings 3.20 and 3.21.

Listing 3.20	Generic example of *for* loop execution with *range(initial_value, end_value, step_value)* function

```
for counter in range(1, 9, 2):
    print("Counter Value", counter)
```

Output of Listing 3.20:

```
Counter Value 1
Counter Value 3
Counter Value 5
Counter Value 7
```

Listing 3.21	Generic example of *for* loop execution (value decrementing) with *range(initial_value, end_value, step_value)* function

```
for counter in range(10, 1, -2):
    print("Counter Value", counter)
```

Output of Listing 3.21:

```
Counter Value 10
Counter Value 8
Counter Value 6
Counter Value 4
Counter Value 2
```

Listings 3.22 and 3.23 demonstrate the usage of the *for* loop in agricultural scenarios.

Listing 3.22	Program demonstrating the use of *for* loop in an agricultural context

```
'''
=================================================================
Python program for agriculturists to calculate the crop yield
over the next five years, increasing each year by 5% on the
initial given yield.
=================================================================
'''
initial_crop_yield = int(input("Enter Initial Yield (kg): "))
growth_rate = 0.05

for year in range(1, 6):
    yield = initial_crop_yield * (1 + growth_rate) ** year
    print(f"Total Yield = {yield:.2f} kg in year {year}")
```

Sample Output of Listing 3.22:

Enter Initial Yield (in kg): 400
Yield = 420.00 kg in year 1
Total Yield = 441.00 kg in year 2
Total Yield = 463.05 kg in year 3
Total Yield = 486.20 kg in year 4
Total Yield = 510.51 kg in year 5

Listing 3.23	Program demonstrating the use of ***for*** loop in an agricultural context

```
'''
=================================================================
Python program calculating the total pesticide cost for farmers
who apply pesticide 3 times in a growing season. The estimated
increase in pesticide cost is $25 each time on the first
treatment cost.
=================================================================
'''
```

```
initial_cost = int(input("Enter first treatment cost: "))
total_treatments = int(input("Enter total treatments: ")) + 1
increment = 25 # in dollars

for treatment_num in range(1, total_treatments):
    cost = initial_cost + (increment * (treatment_num - 1))
    print("Treatment", treatment_num, ":", "$", cost)
```

Sample Output of Listing 3.23:

Enter first treatment cost: 50
Enter total treatments: 3
Treatment 1 : $ 50
Treatment 2 : $ 75
Treatment 3 : $ 100

Like the *nested while* loop, a *nested for* loop in Python is a *for* loop in the body of another *for* loop, and each iteration of the outer *for* loop requires all iterations of the inner *for* loop to be completed. Listing 3.24 demonstrates the use of a *nested for loop* in an agricultural context.

Listing 3.24	Program demonstrating the use of ***nested for*** loop in an agricultural context

```
'''
Python program for a plant breeder who is interested in finding
unique crossbreeding combinations between multiple crop
varieties.
'''

total_varieties = 4

# Taking input for plant varieties one by one
crop_variety1 = input("Enter name of 1st crop variety: ")
crop_variety2 = input("Enter name of 2nd crop variety: ")
crop_variety3 = input("Enter name of 3rd crop variety: ")
crop_variety4 = input("Enter name of 4th crop variety: ")

print("Unique crossbreeding combinations")

# nested for loops to create unique crossbreeding combinations
for v1 in range(1, total_varieties + 1):
    for v2 in range(v1 + 1, total_varieties + 1):
        crop_v1 = eval(f"crop_variety{v1}")
        crop_v2 = eval(f"crop_variety{v2}")
        print(f"{crop_v1} × {crop_v2}")
```

Sample Output of Listing 3.24:

Enter name of 1st crop variety: A
Enter name of 2nd crop variety: B
Enter name of 3rd crop variety: C
Enter name of 4th crop variety: D
Unique crossbreeding combinations
A × B
A × C
A × D
B × C
B × D
C × D

Explanation:
After declaring and initializing the total number of varieties, the program prompts the user to enter the name of the four crop varieties. Next, the outer *for* loop iterates to execute the inner *for* loop. The inner *for* loop executes the three statements related to the dynamic full name creation of varieties along with displaying the unique crossbreeding combinations. Calculate the cost of fertilizer application and pesticide spray for each farm. For each iteration of outer *for* loop, inner loop completes its execution and in each execution it prints unique crossbreeding combinations.

Note: An important use of the for loop is to do processing on data that is stored in Python Sequential Data Structures (also known as Compound Data Types). For Python compound data types, a single variable can hold multiple data items of the same or different data types one after the other (in an organized way). The general syntax of the for loop to deal with data stored in these sequential data structures is given as follows:

```
for variable_name in Python_compound_data_structure
        Statement(s) # for loop body
```

Common Python sequential data structures are Lists, Sets, Tuples, and Dictionaries. The use of a for loop with these Python sequential data structures has been elaborated in Chap. 5.

3.6 Jump Flow Control Statements

In Python, Jump flow control statements disrupt the program's normal execution flow by transferring control to another part of the program. There are four types of Jump flow control statements in Python, i.e., *break*, *continue*, *pass*, and *return*.

3.6.1 *The* Break *Statement*

In Python, the *break* statement stops the normal flow of computer programs. Typically, it is used in loops, where it is required to stop the ongoing iterations and shift the control out of the loop immediately when a certain condition is met, as shown in Listings 3.25 and 3.26.

Listing 3.25	Program to illustrate the use of ***break*** statement in ***while*** loop

```
counter = 10
while counter < 20:
  print("Counter = ", counter)
  if (counter%5) == 4:
     print("Condition True. Breaking out of the loop.")
     break # Terminate and exit the loop immediately
  counter += 1
print("Loop ended.")
```

Output of Listing 3.25:

```
Counter = 10
Counter = 11
Counter = 12
Counter = 13
Counter = 14
Condition True. Breaking out of the loop.
Loop ended.
```

Listing 3.26	Program to illustrate the use of ***break*** statement in ***for*** loop

```
for counter in range(1, 100):
    print("Value of counter = ", counter)
    if (counter / 3) == 2:
        print("Condition True. Breaking out of the loop.")
        break # Terminate and exit the loop immediately
print("Loop ended.")
```

Output of Listing 3.26:

```
Value of counter = 1
Value of counter = 2
Value of counter = 3
Value of counter = 4
Value of counter = 5
Value of counter = 6
Condition True. Breaking out of the loop.
Loop ended.
```

3.6.2 *The Continue Statement*

Programmers use the continue statement to skip code inside the loops for the current iteration and move on to the next iteration. Listings 3.27 and 3.28 show the use of the *continue* statement in *while* and *for* loops, respectively.

Listing 3.27	Program to illustrate the use of *continue* statement in *while* loop

```
counter = 0
while counter < 10:
  counter += 1
  '''
  The continue statement below will skip the iteration of the
  while loop to print the value of the counter when the counter
  value is between 2 and 8.
  '''
  if counter >= 3 and counter <= 7:
    continue
  print("Counter value = ", counter)
print("Loop ends")
```

Output of Listing 3.27:

```
Counter value = 1
Counter value = 2
Counter value = 8
Counter value = 9
Counter value = 10
Loop ends
```

Listing 3.28	Program to illustrate the use of *continue* statement in ***for*** loop

```
for counter in range(0, 10):
  '''
  The continue statement below will skip the iteration of for
  loop to print the value of the counter when the counter value
  is greater than 3 and less than 7.
  '''
  if(counter >= 3 and counter <=7):
    continue
  print("Counter Value = ", counter)
```

Output of Listing 3.28:

```
Counter Value = 0
Counter Value = 1
Counter Value = 2
Counter Value = 8
Counter Value = 9
```

3.6.3 *The* Pass *Statement*

The *pass* statement in Python is known as a null operation statement and is typically used as a placeholder in various Python programming language constructs, i.e., conditional structures, loops, functions, and classes. It is often utilized in situations where a logical block of code or set of instructions is required to be placed in the future, but the programmer at the moment has no instructions to execute. The use of the *pass* statement in a *for* loop has been illustrated with the help of an example in Listing 3.29.

Listing 3.29	Program to illustrate the use of *pass* statement

```
'''
=================================================================
Python program to check the operational status of sensors
deployed in an agricultural farm.
=================================================================
```

```
'''
print("Field sensor status checking")
# Check status of all sensors with IDs (1 to 4)
for sensor_id in range(1, 5):
    print("Checking status for Sensor ID: ", sensor_id)
    # suppose sensors with ID 3 and ID 5 are nonfunctional
    if sensor_id == 2 or sensor_id == 3:
      '''
      Use pass as placeholder to implement future functionality
      for non-operational sensors
      '''
      pass
    else:
      print("Sensor with ID", sensor_id, "working properly")
```

Output of Listing 3.29:

```
Field sensor status checking
Checking status for Sensor ID: 1
Sensor with ID 1 working properly
Checking status for Sensor ID: 2
Checking status for Sensor ID: 3
Checking status for Sensor ID: 4
Sensor with ID 4 working properly
```

3.6.4 The **Return** *statement*

In Python, a *return* statement is a type of jump statement that is used to finish a function's (or method's) execution while returning control to the calling function (or method). Moreover, a value can be returned to the calling function (or method) with this statement. The details of this statement have been discussed in Chap. 4.

3.7 Exercises

Problem 3.1 A digital smart greenhouse system is required to be implemented to check and adjust soil moisture within the greenhouse on a daily basis. The criteria to adjust the soil moisture level in the greenhouse are given as follows:

If the moisture level is 20% less than the minimum moisture threshold (i.e., 70% Volumetric Water Content (VWC)), then it opens the irrigation system for 1 hour every day. This 1-hour watering of plants in the greenhouse increases the moisture level by 25% of the current moisture level. Write a Python program using a *while* loop to explain the proper working of this system.

Problem 3.2 Write a Python program for an automated sprinkler system that is required to be deployed in a garden that is used to water plants every day for 1 hour (30 min for the right side of the garden and 30 min for the left side of the garden).

Problem 3.3 A smart pest monitoring system is equipped with a camera that is attached to an image processing module. The camera is supposed to take a photo of plants in a specific area of an agricultural field every day, which is sent to the image processing module to count the number of pests on plants in that photo. Write a Python program for a submodule of the image processing system that takes the pest count as input and advises the farmer whether to spray pesticide, based on whether the pest count exceeds a threshold of 15.

Problem 3.4 Write a Python program to assist farmers with pest management recommendations. The system should take into account three key inputs from the user: the type of crop being cultivated, the type of pest affecting the crop, and the severity level of the infestation.

- The input for crop type can either be wheat or rice.
- Depending on the selected crop, the system should present relevant pest options, i.e.,
  - for Wheat, the pest options are aphids or armyworm,
  - for Rice, the options are stem borer or leaf folder.
- The input regarding the severity level of the infestation can be low, medium, or high.

Based on the combination of these inputs, the system will generate tailored recommendation advice for farmers. For example, the program will take the user's input and provide advice based on these rules.

- if the selected crop is wheat and the pest is aphids, the system will suggest
  - "Crop monitoring" recommendation advice for a low severity level pest infestation area,
  - "Use suitable biological control" recommendation advice for a medium severity level area, and
  - "Spray an appropriate pesticide immediately" recommendation message in case of high severity infestation.

- If the selected crop is wheat and the pest is armyworms, the system will suggest
  - "Use suitable biological control" for both low and medium severity levels and
  - "Spray an appropriate pesticide immediately" for high severity level.
- In the case of rice, if the pest is Stem borer, the recommendation will always be to "spray an appropriate pesticide immediately," regardless of the severity level.
- However, for Rice affected by Leaf folder, the system will consistently send the "Use suitable biological control" message to the farmer (irrespective of the level of infestation severity).

Chapter 4
Python Functions, Methods, and Modules

4.1 Python Function and Method

A function (or method) in programming refers to a reusable segment of code designed to perform a specific task. Functions are executed when invoked or called within another function. A calling (or caller) function is a function that initiates the execution of another function, while the called function is the function that is invoked or executed. Upon the completion of the called function's execution, control is returned to the calling function. In general, based on the same purpose and working philosophy, the terms function and method are often used interchangeably; however, in Python, some subtle differences exist, as mentioned in Table 4.1.

Note: Although the differences between a function and a method are summarized in Table 4.1, *the terms will be used interchangeably throughout the rest of this book.*

4.2 Benefits of Using Function/Method

Functions (and methods) offer several benefits in computer programming, i.e.:

- ***Modularity***: Breaking a (big) program into smaller parts that are logically related and easier to organize.
- ***Readability***: Make programs clearer and more readable.
- ***Code reusability***: Reduce redundancy as the same block of code or set of instructions can be reused multiple times.
- ***Easy debugging***: Programs written using functions are easier to debug and test, as issues can be isolated to specific function(s) without affecting the rest of the program.
- ***Scalability***: Allow programs to grow without disturbing the existing codebase.

M. A. Iqbal, *Python for Agriculturists*,
https://doi.org/10.1007/978-3-032-01442-9_4

Table 4.1 Differences between Python function and Python method

Parameter	Function	Method
Definition	Functions are not defined within any class	Methods are always defined within a class
Invocation	Functions are invoked simply by their name	Methods are invoked on an object
Association	Not bound to any class objects	Always bound with class object(s)
Operation	Operates on the data that is passed as argument(s)	Operates on the data of the object it associates with along with the data that is passed as argument(s)
Implicit arguments	Requires all arguments to be explicitly passed. So, Functions do not require any self-argument	Takes the object it is called on as an implicit first argument (conventionally named *self*)

– ***Collaboration***: In team projects, functions allow developers to work on different parts of the program simultaneously, improving productivity and collaboration.

4.3 Python Functions and Control Flow

Control flow is the sequence of program execution. The control flow of a program begins with the first statement of the main program but does not always follow a strict sequential order. For example, control flow skips over lines when a conditional or repetitive statement is executed or when execution reaches a function call. The control flow in cases of conditional and repetitive statements has already been elaborated in Chap. 3. In the case of a function call, the control jumps from the calling function statement to the position where the function is defined to execute the function statement(s) written in the function body. After a function completes its execution, control returns to the point in the program where the function was called, and the remaining statements continue to be executed.

4.4 Types of Functions in Python

User-defined functions and built-in functions are the two main types of functions in Python.

4.4.1 User-Defined Functions

User-defined functions are not predefined functions and are created by programmers to fulfill their specific needs. The basics of user-defined functions have been discussed below.

4.4.1.1 Defining a User-Defined Function

In Python, it is essential to define a function before using it in the program. Essentially, the function definition can be divided into two parts, i.e., function header and function body. The generic structure of the function definition is given as follows:

```
def function_name (optional_parameters):
    '''
    Documentation string describing the function's purpose along
    with mentioned parameter(s) and function return type.
    :param optional_parameters:
    :return: optional return statement
    '''

    Statement(s)
    optional_explicit_return_statement
```

The function header consists of the *def* keyword, function name, and optional parameters in parenthesis, followed by a colon (:) at the end. Here, *def* (a short form of define) is a Python keyword that is used to define a function. Function names should be meaningful and convey the meanings of the function's behavior or output. Moreover, for function names, you have to consider all the basic rules of naming identifiers that have already been described in Sect. 2.3.1 of Chap. 2. Optional parameter(s) inside parenthesis of the function definition (also known as formal parameter(s)) is/are the placeholder(s) for the actual parameter(s). It is important to mention here that default values for formal parameters can be mentioned in the function definition, and these values are considered for processing if no arguments are passed to the calling function. The syntax of using default parameter values in the function definition is given as follows:

```
def function_name(parameter1=value1,…,parameterN=valueN):
```

Actual parameter(s), also known as argument(s), are the value(s) passed to the calling function when that function is called or invoked.

The indented body of a function should optionally start with comments known as documentation string explaining the purpose of the function, followed by the set of instructions or statements that actually define its behavior. After executing the set of instructions, the function either immediately returns control to the calling function or can optionally pass value(s) back to the caller using the explicit *return* statement.

4.4.1.2 User-Defined Function Calling

Function call or function invocation in Python means executing a predefined function. Function calls in Python are possible by using a function name followed by parenthesis. The generic structure of the function call in Python is

```
function_name(optional_argument(s)_passing)
```

If the function does not take any arguments, you can call it using empty parenthesis with the function name; however, for functions taking arguments, you can pass arguments inside these parentheses. There are two ways in Python to pass arguments to functions, i.e., *position-based argument(s) passing* and *keyword-based argument(s) passing*. In the position-based category, the arguments are passed in the same order of parameters as mentioned in the function definition. On the other hand, in the case of keyword-based argument passing, parameter names are used to pass the arguments during the function call, and the order of these passed arguments can be different from the order of parameters mentioned in the function definition. The syntax of passing arguments in both cases (positional and keyword arguments) is given below.

Syntax for position-based argument passing

```
function_name(value1,value2,value3, …, valueN)
```

Syntax for keyword-based argument passing

```
function_name(paramName1=value1, …, paramNameN=valueN)
```

These various ways of defining function and function calls are demonstrated in Listings 4.1–4.7.

Listing 4.1	Example demonstrating the defining and calling of a user-defined function in Python without any parameters

```
# Function Definition
def welcome_message():
    '''
     A simple user-defined function in Python (without any
     parameters and explicit return statement) to print the
     Welcome message
    :param no parameter
    :return: None
    '''
```

```
    print("Welcome to Digital Agriculture Function")

# Program execution starting point
print("Before Function Call")

# Function call
welcome_message()

print("After Function Call")
print("Program Execution Completed!")
```

Output of Listing 4.1:

Before Function Call
Welcome to Digital Agriculture Function
After Function Call
Program Execution Completed!

Explanation:
In Listing 4.1, the code execution begins with the first executable print statement outside the defined function in the program, i.e., ***print("Before Function Call")***. After printing "***Before Function Call***," the interpreter executes the function call statement, i.e., ***welcome_message()***, and control is transferred to the ***welcome_message()*** function. Within this function, the print statement outputs "***Welcome to Digital Agriculture Function***." Once the function execution is completed, control returns to the original calling location, where the next print statement is executed, displaying the "***After Function Call***." Finally, the last print statement outputs ***Program Execution Completed!***, indicating the end of the program.

Listing 4.2	Example demonstrating the defining and calling of a user-defined Python function with one parameter

```
def welcome_message(ag_domain):
    """
     A simple user-defined function in Python to accept one
     parameter to print the passed argument to this function
    :param ag_domain
    :return: None
    """
    print("Welcome to " + ag_domain)

# Program execution starting point
print("Before Function Call")

# Function call
welcome_message("Smart Farming")
welcome_message("Precision Farming")
print("After Function Call")

print("Program Execution Completed!")
```

Output of Listing 4.2:

Before Function Call
Welcome to Smart Farming
Welcome to Precision Farming
After Function Call
Program Execution Completed!

Explanation:
In Listing 4.2, the program execution begins with the first executable print statement outside the defined function in the program, i.e., ***print("Before Function Call")***. After displaying "***Before Function Call***" to the console, the interpreter executes the next two function call statements, i.e., ***welcome_message()***, while passing different string parameters each time (such as "***Smart Farming***" the first time and "***Precision Farming***" the second time). Each time, the control is transferred to the ***welcome_message() function***. The first few lines of the ***welcome_message()*** function are the docstring (comments) that describe the overall working of the function. For each call, the output is different depending on the passed argument. For the first function call, the print statement outputs "***Welcome to Smart Farming***," and for the second function call, the same function prints "***Welcome to Precision Farming***." Once both function executions are completed, the control returns to the original calling location, where the next print statement is executed, displaying the "***After Function Call***." Finally, the last print statement outputs ***Program Execution Completed!***, indicating the end of the program.

Listing 4.3	Example demonstrating that arguments can be passed to the Python function positionally or using keywords

```
def crop_info(crop, year, crop_yield):
    '''
     A simple user-defined Python function with three parameters
     that can be called by passing arguments positionally and by
     keywords
    :param crop
    :param year
    :param crop_yield
    :return: None
    '''
    print(crop, "yield in", year, "is", crop_yield, "tonnes/acre")

# Program execution starting point
print("Before Function Call")

# Function call by passing arguments positionally
crop_info("Wheat", 2023, 8.5)

# Function call by passing arguments by keywords
crop_info(crop_yield = 9.1, crop = "Wheat", year = 2024)

print("After Function Call")

print("Program Execution Completed!")
```

Output of Listing 4.3:

Before Function Call
Wheat yield in 2023 is 8.5 tons/acre
Wheat yield in 2024 is 9.1 tons/acre
After Function Call
Program Execution Completed!

Explanation:
In Listing 4.3, the program execution begins with the first executable print statement outside the defined function in the program, i.e., ***print("Before Function Call")***. The first output displayed on the console is "***Before Function Call***." The interpreter then proceeds to execute two function calls. The first function call uses positional argument passing, and the other uses keyword-based argument passing. During the first function call, the function is invoked with three parameter values passed positionally. The function executes and prints:
"***Wheat yield in 2023 is 8.5 tons/acre.***"
Once the function completes its execution, control returns to the original calling location, and the program moves to the next statement, which is another function call. In the second function call, arguments are passed using keyword-based argument passing. Although the parameters are not provided in the same order as defined in the function, each argument is correctly assigned to the respective parameter based on its name. As a result, the function executes and prints:
"***Wheat yield in 2024 is 9.1 tons/acre.***"
After the function completes its execution, control once again returns to the calling location, and the next print statement is executed, displaying "***After Function Call***." Finally, the last print statement outputs ***Program Execution Completed!***, indicating the end of the program.

Listing 4.4	Example demonstrating ways to call a Python function having default arguments

```
def irrigation_time(crop = "Maize", soil_moisture = 35):
    '''
     A Python function to facilitate farmers with information
     about irrigation time while considering crop type and
     current soil moisture. The default crop is 'Maize' and the
     default soil moisture is 35%.

    :param crop
    :param soil_moisture
    :return: None
    '''
    if soil_moisture < 30:
        if crop == "Wheat":
            time = 25 # minutes
        elif crop == "Maize":
            time = 30 # minutes
        else:
            time = 35 # minutes for other crops
        print("Irrigate the", crop, "field for", time, "minutes.")
    else:
```

```
        print("Irrigation is not needed for", crop,)

# Program execution starting point
print("Starting Program Execution")

# Function calling in different ways
# Defaults to 'Maize' and soil moisture 35%
irrigation_time()

# For Wheat with 20% soil moisture
irrigation_time("Wheat", 20)

# For Wheat with default 35% soil moisture
irrigation_time("Wheat")

# Defaults to Maize with 22% soil moisture
irrigation_time(soil_moisture=22)

print("Program Execution Completed!")
```

Output of Listing 4.4:

Starting program execution
Irrigation is not needed for Maize
Irrigate the Wheat field for 25 minutes.
Irrigation is not needed for Wheat
Irrigate the Maize field for 30 minutes.
Program Execution Completed!

Explanation:
In Listing 4.4, the program execution begins with the first executable print statement outside the defined function in the program, i.e., ***print("Starting program execution")***. Therefore, the first output displayed on the console is "***Starting Program Execution***." The interpreter then proceeds to execute the function calls. The first function call is without passing any arguments. For this function call, default parameter values (i.e., crop name is Maize and soil moisture is 35%) in the function definition will be considered for function logic, and hence the output will be:
"***Irrigation is not needed for Maize***".
Once the function completes its execution, control returns to the original calling location, and the program moves to the next statement, which is another function call. In the second function call, position-based arguments are passed explicitly. Explicitly passed arguments have priority over default arguments, and therefore for this function call, the output will be:
"***Irrigate the Wheat field for 25 min***".
Once the function completes its execution this time, control returns to the original calling location, and the program moves to the next statement, which is another function call. For this third time function call, only the crop name parameter has been passed explicitly and has priority over default (crop) function parameter. As no value is passed for the second parameter this time, the function will use its default value, resulting in an output of
"***Irrigation is not needed for Wheat***".
Once the function completes its execution this time, control returns to the original calling location, and the program moves to the next statement, which is another function call. For this fourth time function call, only the value for the second function parameter is passed explicitly (as a keyword-based argument) and has priority over the default soil moisture parameter value. Considering this state, for this function call, the output will be:
"***Irrigate the Maize field for 30 min***".
After the completion of function execution this time, the control once again returns to the calling location, and the next print statement is executed, displaying "***Program Execution Completed!***".

4.4.1.3 Types of Argument Passing in User-Defined Functions

In Python, there are two ways to pass arguments, i.e., *Pass-by-Value* and *Pass-by-Reference*. To grasp the distinction between *Pass-by-Value* and *Pass-by-Reference*, it is important to first understand two key concepts of the Python language: objects and the mutability of objects. Although a detailed discussion of these concepts falls outside the scope of this book, a brief explanation is provided to clarify the two approaches of passing arguments to functions. In Python, it is important to understand that it is an object-oriented language, and it is often stated that "everything in Python is an object," including numbers, strings, functions, and modules. This means that all these elements are instances of some class, even if the programmer is not explicitly writing object-oriented code. Another key aspect to consider is that some objects in Python are mutable (can be modified after creation), while others are immutable (cannot be modified once created). Examples of immutable objects are integer, float, string, Tuple, and so on, and examples of mutable objects are Sets, Dictionaries, and Lists.

Note: Details about List, Set, Dictionary, and Tuple are covered in Chap. 5. *However, in this chapter, a List (i.e., one of Python's compound data types and allows multiple values to be stored within a single variable) is used specifically to illustrate the difference between Pass-by-Value and Pass-by-Reference in argument passing.*

4.4.1.3.1 Argument Passing by Value

When immutable objects are passed to a function as arguments, the function creates a separate copy of that value. Within the function, all the processing is done using the newly created copy, and the value stored in the original variable remains unchanged outside the function because these objects cannot be modified, as shown in Listing 4.5. Figure 4.1 provides a pictorial explanation of the code used in Listing 4.5.

Listing 4.5	Example demonstrating the execution of a function when arguments are passed by value

```
def update_yield(crop_yield):
    '''
    A Python function to modify the value of a crop yield with
    the updated value that is passed as an argument.
    :param crop_yield:
    :return: None
    '''
    print("In function: Yield =", crop_yield, "tons")
```

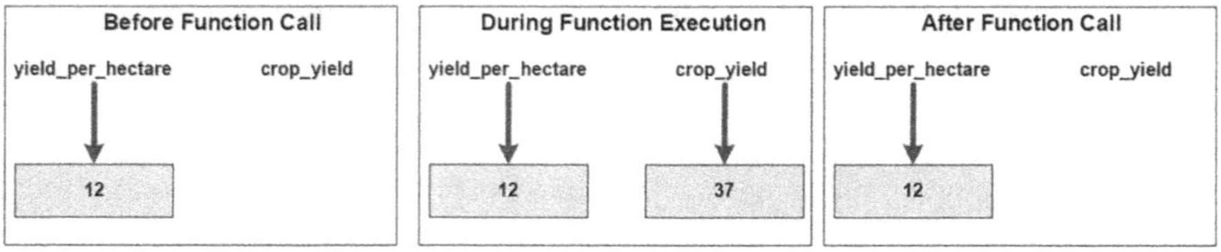

Fig. 4.1 Pass-by-value representation

```
    crop_yield += 25 # This modifies a local copy
    print("Updated Yield in function=", crop_yield, "tons")

# Initial crop yield
yield_per_ha = 12 #tonnes
print("Yield before function call:", yield_per_ha, "tons")
update_yield(yield_per_ha)
print("Yield after function call:", yield_per_ha, "tons")
```

Output of Listing 4.5:

Yield before function call 12 tons
In Function: Yield= 12 tons
Updated Yield in function= 37 tons
Yield after function call: 12 tons

Explanation:
From the output of Listing 4.5, it becomes evident that crop yield is modified inside the function, but outside the function it remains the same. This is due to the reason that for the passed immutable argument, a separate copy local to the function is created for further processing, as shown in Fig. 4.1. After the completion of function execution, the local copy is destroyed. This is the reason that any changes in the passed argument(s) are not reflected outside the function body.

4.4.1.3.2 Argument Passing by Reference

When a custom object or an object of mutable compound data types, i.e., List, Dictionary, or Set is passed to a Python function, then instead of making copies for further processing, Python allows modifications to the (originally) passed object. Therefore, modifications made inside the function on the received parameter are reflected outside. In other words, this is because in this case the function operates on a reference to the same (original) object as shown in Listing 4.6. Figure 4.2 provides a pictorial explanation of the code used in Listing 4.6.

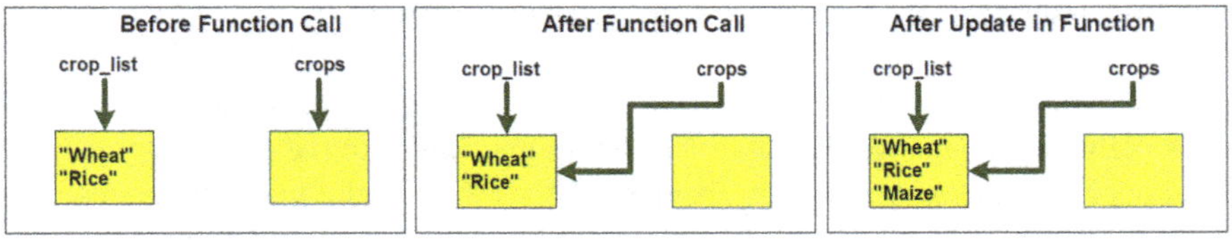

Fig. 4.2 *Pass-by-reference* representation

Listing 4.6	Example demonstrating the execution of a function where arguments are passed by reference

```
def update_crop_list(crops):
    '''
    A Python function to modify the list of crops with the
    updated value that is passed as argument.
    :param crops:
    :return: None
    '''
    print("In function: crop list before update", crops)

    # Original list modification by appending new item
    crops.append("Maize")
    print("In function: crop list after update", crops)

# List of crops
crop_list = ["Wheat", "Rice"]
print("Crop list before function call", crop_list)
update_crop_list(crop_list)
print("Crop list after function call:", crop_list)
```

Output of Listing 4.6:

Crop list before function call ['Wheat', 'Rice']
In function: crop list before update ['Wheat', 'Rice']
In function: crop list after update ['Wheat', 'Rice', 'Maize']
Crop list after function call: ['Wheat', 'Rice', 'Maize']

Explanation:
The output of Listing 4.6 demonstrates that the (mutable) crop list (object) is modified within the function, and these changes are also reflected in the original list outside the function. This happens because rather than creating a separate copy of the mutable object for further processing, the Python function works with the same original mutable object that was passed as an argument. As a result, any modifications made within the function are also visible outside its scope, as illustrated in Fig. 4.2.

4.4.1.4 User-Defined Functions with Return Statement

Python functions without explicit *return* statements perform some actions (and optionally display the result) but do not return any values to the caller, as you have seen in previous examples of this chapter. On the other hand, Python functions with explicit return statements typically after performing some computation or processing, return value(s) to the caller. Listings 4.7–4.9 demonstrate the use of functions with explicit return statements.

Listing 4.7	Python function returning the calculated amount of fertilizer required for a given area

```
def fertilizer_amount(area, rate):
    '''
    Python function that returns the required amount of
    fertilizer (in kg) for a given area (in hectares)
    :param area:
    :param rate:
    :return: area x rate
    '''
    return area * rate

total_area = 1.5 # in hectares
rate_per_ha = 35 # kg per hectare

# Calling the fertilizer_amount() function
fertilizer_req = fertilizer_amount(total_area, rate_per_ha)

# Display the returned value
print("Required Fertilizer Amount is", fertilizer_req, "kg.")
```

Output of Listing 4.7:

Required Fertilizer Amount is 52.5 kg.

Explanation:
In Listing 4.7, the function *fertilizer_amount()* calculates and returns the total fertilizer required for the fertilization of a particular area, which is assigned to the variable *fertilizer_needed* that can be used later in the program.

Listing 4.8	Python function evaluating and returning the growth stage of crop plants based on their height

```
def crop_growth_stage(height):
    '''
    Function to determine and return the crop's growth stage
    based on the given crop height. The criteria are given
    as under:
  - Seedling stage if the crop height is less than 10 cm
  - Vegetative stage if the crop height is between 10cm and 50cm
  - Flowering stage if the crop height is between 40cm and 100cm
  - Mature stage if the crop height is above 100cm
  - :param height
  - :return: String
    '''
    if height < 10:
        return "Seedling Stage"
    elif height < 50:
        return "Vegetative Stage"
    elif height < 100:
        return "Flowering Stage"
    else:
        return "Mature Stage"

# Taking user input for crop height in cm
crop_height = eval(input("Enter crop height in cm: "))

# calling the crop_growth_stage() function
growth_stage = crop_growth_stage(crop_height)

# Display the returned value
print("The crop is in the", growth_stage)
```

Sample Output of Listing 4.8:

Enter crop height in cm: 15
The crop is in the Vegetative Stage

Description
Crop height is passed to the function, and as soon as a condition is met, the corresponding return statement is executed, and the function exits.

Listing 4.9	Python function evaluating and returning both the ideal height and number of days to harvest a crop

```
def harvest_limit(crop_name):
    '''
    Function to provide information about ideal crop height and
    number of days to harvest a crop. The criteria are given
    as under:
    - For wheat crop, ideal height is 69 cm and 190 days
      to harvest
    - for Rice crop, ideal height is 110 cm, and 125 days
      to harvest
    :param crop_name:
    :return: height and number of days to harvest
    '''

    if crop_name == "Wheat":
        height = 69
        days_to_harvest = 190
        return height, days_to_harvest
    elif crop_name == "Rice":
        height = 110
        days_to_harvest = 125
        return height, days_to_harvest

# Calling the function
crop_height, harvest_days = harvest_limit("Rice")

# Display the returned values
print("Harvesting Height (cm):", crop_height)
print("Days to Harvest:", harvest_days)
```

Sample Output of Listing 4.9:

Harvesting Height (cm): 110
Days to Harvest: 125

Description
Crop name is passed to the function, and as soon as a condition is met within the function, the corresponding return statement is executed, and multiple values (i.e., ideal crop height and days to harvest) are returned to the calling function. The returned values are stored in two variables (i.e., *crop_height* and *total_days_to_harvest*) for further processing in the main program (outside the function's scope).

4.4.1.5 User-Defined Functions and Variable Scope

The scope of a variable in a programming language defines the accessibility of that variable in different parts of the computer program. There are four types of variable scope (i.e., *Local*, *Global*, *Enclosing*, and *Built-in*) in Python. The comprehension of these variable scopes ensures the writing of cleaner and error-free code that avoids unintended modifications. Below is the explanation of all these scopes.

4.4.1.5.1 Local Scope

Variables defined in a function are accessible only within that function and that is why their scope is considered as Local Scope (to that function) as demonstrated in Listing 4.10.

Listing 4.10	Python program to demonstrate the scope of global variables

```
def fieldA_area():
    lengthA = 30
    widthA = 40
    return lengthA * widthA

def fieldB_area():
    lengthB = 10
    widthB = 20
    return lengthB * widthB

print(lengthA * widthA + lengthB * widthB)
```

Output of Listing 4.10:

NameError: name "lengthA" is not defined

Explanation
If you execute this program, then because of the use of variable names lengthA, widthA, lengthB, and widthB in the print statement (outside both functions), you will encounter the following error message.

NameError: name 'lengthA' is not defined

This is because these variables are defined within the functions. The scope of these variables is limited or local to these functions only and is not accessible outside these functions.

If you want to execute this program without error, then replace the mentioned print statement with the print statement given as follows:
print(fieldA_area() + fieldB_area())

4.4.1.5.2 Global Scope

Variables defined outside any function (or other coding block) are accessible from anywhere in the code and that is why their scope is considered as Global Scope (to that program). The accessibility of global variables within a Python program is shown in Listing 4.11.

Listing 4.11	Python program to demonstrate the accessibility of global variables

```
# Defining the global variable named area
area = 0

def fieldA_area():
    lengthA = 30 # Local variable
    widthA = 40 # Local variable
    area = lengthA * widthA
    return area # Global Variable in fieldA_area() function

def fieldB_area():
    lengthB = 10 # Local variable
    widthB = 20 # Local variable
    area = lengthB * widthB
    return area # Global Variable in fieldB_area() function

# Use of global variable outside functions
area = fieldA_area() + fieldB_area()
print("Total Area =", area)
```

Output of Listing 4.11:

Total Area = 1400

Explanation
From this Python code, it becomes evident that a global variable remains accessible to all parts of the program. In this example, the global variable area remains accessible both inside and outside the user-defined functions.

If a local variable has the same name as the global variable, then the local variable dominates over the global variable in that function, which is known as *shadowing of the global variable*.

Listing 4.12	Python program to demonstrate the shadowing of global variables

```
area = 0
print("Area at the start of the program is", area)
def fieldA_area():
    lengthA = 30 # Local Variable
    widthA = 40 # Local Variable

    # Local variable area shadows the global variable
    area = lengthA * widthA
    print("Area inside fieldA function", area)

def fieldB_area():
    lengthB = 10 # Local Variable
    widthB = 20 # Local Variable
    area = lengthB * widthB

    # Local variable area shadows the global variable
    print("Area inside fieldB function", area)

fieldA_area()
fieldB_area()
print("Area at the end of the program is", area)
```

Output of Listing 4.12:

Area at the start of the program is 0
Area inside fieldA function 1200
Area inside fieldB function 200
Area at the end of the program is 0

Explanation

This Python code demonstrates that a local variable within a function with the same name as a global variable dominates within that function. At the beginning of the program, the variable area is initialized globally with a value of 0. However, when the functions fieldA_area() and fieldB_area() define their own local variables named area, these local variables overshadow the global variable within their respective function scopes. As a result, the value of area variables inside these functions is different from the globally assigned value. Outside these functions, the global area remains unchanged at 0 since modifications made to the local variables within the functions do not affect the global variable.

The *global* keyword of Python can be used to modify global variables inside a function, as shown in Listing 4.13.

Listing 4.13 | Python program demonstrating the avoidance of global variable shadowing

```
# Declaring and initializing the Global Variable
total_arid_land = 100 # Hectares

def update_aridland_area():
    # use of global keyword to avoid shadowing
    global total_arid_land
    total_arid_land += 120

update_aridland_area()
print("Updated total arid land:", total_arid_land, "Hectares")
```

Output of Listing 4.13:

Updated total land: 220 Hectares

Explanation

The Python code in Listing 4.13 illustrates the use of global variables within a user-defined function while avoiding the shadowing effect. By using the *global* keyword, the Python program is instructed to access the globally declared variable instead of creating a new local variable within the function's scope. This use of the global keyword within the function ensures that any modifications made to the variable inside the function directly affect the globally defined variable.

4.4.1.5.3 Enclosing (Nonlocal) Scope

Variables in an enclosing function (outer function) can be accessed by an inner function using the *nonlocal* keyword, as shown in Listing 4.14.

Listing 4.14 | Python program demonstrating the use of enclosing (or nonlocal keyword) use

```
def field_perimeter():
    '''
    Outer function calculating perimeter of an agricultural field
    using two variables length and width
    :return: None
    '''
    length = 30 # Outer Function Variable
    width = 40 # Outer Function Variable
    print("Field Perimeter =", (2 * (length + width)))
    def field_area():
        '''
```

```
        Inner function using the same two variables length and
        width that are already defined in outer function to
        calculate the area of an agricultural field.
        :return: None
        '''
        nonlocal length # Outer Function Variable
        nonlocal width # Outer Function Variable
        print("Field Area =", length * width)

    field_area() # Calling Inner Function
field_perimeter() # Calling Outer Function
```

Output of Listing 4.14:

Field Perimeter = 140
Field Area = 1200

Explanation

The use of the nonlocal Python keyword in Listing 4.14 allows the accessibility of outer function variables within the inner function.

4.4.1.5.4 Built-in Scope

Python has various built-in constants (e.g., *False*, *True*, *None*, etc.) and built-in functions (e.g., *len()*, *print()*, *eval()*, *float()*, etc.), which are always available in Python programs.

4.4.2 Built-in Functions

Built-in functions in Python are predefined functions that are designed to perform specific tasks, i.e., displaying program output, string manipulation, mathematical operations, manipulation of Python data structures, and file handling. Built-in functions to deal with file handling operations and manipulate Python data structures (i.e., List, Tuple, Set, Dictionary, etc.) have been discussed in detail in Chaps. 5 and 6, respectively. However, some of the most commonly used built-in Python functions associated with output, string manipulation, and mathematical operations have been discussed below.

4.4.2.1 The *input()* Function

The input function is used to get information from the user. The general syntax of the input function is shown as follows:

```
variable_name = input()
```

or

```
variable_name = input("Enter the appropriate message string here")
```

By default, the input function takes the input as a string and stores it in the variable on the left-hand side of the assignment operator. If you want to convert the type from string to a number, then you can use ***eval()***, ***float()***, or ***int()*** functions.

Examples:

```
crop_name = input("Enter the crop name")
harvesting_year = eval(input("Enter year of harvesting"))
crop_yield = float(input("Enter crop yield"))
crop_area_ID = int(input("Enter area ID"))
```

4.4.2.2 Built-in Output (*print()*) Function

The built-in function *print()* is the most commonly used function that displays the strings (messages) or objects' values to the console. The entire syntax of the Python print() function along with the description of all parameters is given as follows:

```
print(object(s), sep = ' ', end = '\n', file=sys.stdout, flush=False)
```

All parameters of the print() function are optional, with the following semantics:

- Object(s) indicate(s) one or more objects and values to be printed that are converted to strings before printing
- The *sep* parameter specifies how multiple objects passed to the print() function will be separated. The default value of this parameter is ' ' (empty string).
- The *end* parameter specifies the last character to be displayed. The default value of this parameter is '\n' (newline character).

- The *file* parameter is used to redirect the output of the print() function to a file or stream instead of the default *sys.stdout*.
- The Boolean parameter *flush* can take either *True* or *False* Boolean value. When *True*, this parameter forces the output to be written to the file or stream immediately or buffered when set to *False*.

It is important to remember that other than the object(s) parameter, all parameters, i.e., *sep*, *end*, *file*, and *flush* are keyword arguments, and it is compulsory to specify their keywords when there is a need to use these parameters explicitly. There are so many ways to use print statements, but the most commonly used ways have been shown in Listing 4.15.

Listing 4.15	Different ways to use the built-in ***print()*** function

```
print("Start of the Program")
print("=============Section 1=============")
print(100)
print(100 + 200)
print("=============Section 2=============")
crop_name = "Cotton"
print('The crop name is', crop_name)
print("=============Section 3=============")
harvesting_year = 2025
crop_yield = 13
print("Yield for year", harvesting_year, "is", crop_yield)
print("=============Section 4=============")
print('Rice', 'Wheat', 'Maize', 'Oat', sep = ' | ')
print("=============Section 5=============")
print('Integrated Pest Management', '(IPM)', end = '\n')
print("=============Section 6=============")

# new line using sep parameter
print('Weeds', 'Pests', 'Rodents', 'Diseases', sep = '\n')

# new line in the objects parameter
print("=============Section 7=============")
print('Vegegtables \n Fruits')
print("End of the Program")
```

Output of Listing 4.15:

The self-explanatory output of Listing 4.15 is shown below.

```
Start of the Program
=============Section 1=============
100
300
=============Section 2=============
The crop name is Cotton
=============Section 3=============
Yield for year 2025 is 13
=============Section 4=============
Rice | Wheat | Maize | Oat
=============Section 5=============
Integrated Pest Management (IPM)
=============Section 6=============
Weeds
Pests
Rodents
Diseases
=============Section 7=============
Vegegtables
Fruits
End of the Program
```

4.4.2.2.1 Formatting in *print()* Function

Strings and data can be formatted within the print() function using the following three different methods:

- using *format()* function
- using formatted string literals (with *f-strings*)
- old-style string formatting (with *% operator*)

The use of these ways to format output using the *print()* function has been explained with the help of examples in Listing 4.16.

Listing 4.16	Different ways to do formatting in ***print()*** function

```
crop_name1 = "Wheat"
crop_yield1 = 30.9967 #tonne

# Use of format() function
print("1a - Using .format() function")
formatted_string = "Crop Name: {}, Crop Yield: {}".format(crop_
                                              name1, crop_yield1)
```

```
print(formatted_string)

# format() function to round no. to two decimal places.
print("1b - Use format() to round to two decimal places")

formatted_string = "Crop Name: {}, Crop Yield:
                          {:.2f}".format(crop_name1, crop_yield1)

print(formatted_string)

# Using formatted string literals (or with f-strings)
print("2a - Using formatted string literals (with f-strings)")
crop_name2 = "Rice"
crop_yield2 = 35.557
formatted_string = f"Crop Name: {crop_name2}, Crop
Yield:{crop_yield2}"
print(formatted_string)

# Use of f-strings while rounding to two decimal places
print("2b - f-strings with rounding to two decimal places")

formatted_string = f"Crop Name: {crop_name2}, Crop Yield:
                                            {crop_yield2:.2f}"
print(formatted_string)

# Use of f-strings to evaluate expressions
print("2c - Using f-strings to evaluate expression")
print(f"Total Yield {crop_yield1 + crop_yield2}")
'''
Old-style string formatting with the % operator.
It is used to insert and format strings into a template string.
'''
print("3 - Old-style string formatting (with % operator)")
crop_name = "Maize"
crop_yield = 50
formatted_string = "Crop Name: %s, Crop Yield: %d" % (crop_name,
                                                      crop_yield)

print(formatted_string)
```

Output of Listing 4.16:

The self-explanatory output of Listing 4.16 is shown below.

1a - Using .format() function
Crop Name: Wheat, Crop Yield: 30.9967
1b - Use format() to round to two decimal places
Crop Name: Wheat, Crop Yield: 31.00
2a - Using formatted string literals (with f-strings)
Crop Name: Rice, Crop Yield: 35.557
2b - f-strings with rounding to two decimal places
Crop Name: Rice, Crop Yield: 35.56
2c - Using f-strings to evaluate expression
Total Yield 66.5537
3 - Old-style string formatting (with % operator)
Crop Name: Maize, Crop Yield: 50

4.4.2.3 Built-in *(Data) Type Conversion* Functions

Python provides built-in functions for converting data from one type to another. These functions can be used to convert between different data types, such as integers, floats, strings, and even compound data types (discussed in Chap. 5). The use of some of the most commonly used type conversion functions in Python is demonstrated in Listing 4.17.

Listing 4.17	Different ways to do formatting in ***print()*** function

```
# To convert the entered string value into integer data type
sowing_year = int(input("Enter the crop sowing year: "))

# To convert the entered string value into float data type
crop_yield = float(input("Enter the crop yield: "))

# To convert the entered string value to numeric data type
crop_price = eval(input("Enter the crop price: "))

# To convert the specified numerical value to string data type
harveting_month = str(11)

'''
To convert numerical value to boolean data type
- bool() returns false for 0 value
- bool() returns true for any numeric value other than 0 '''
pesticide_needed = bool(0)
fertilizer_required = bool(1)

print("Sowing Year:", sowing_year)
print("Crop Yield:", crop_yield)
print("Crop Price:", crop_price)
print("Harvesting Month:", harveting_month)
print("Pesticide Needed:", pesticide_needed)
print("Fertilizer Needed:", fertilizer_required)
```

Output of Listing 4.17:

The self-explanatory output of Listing 4.17 is shown below.

Enter the crop sowing year: 2025
Enter the crop yield: 61
Enter the crop price: 70
Sowing Year: 2025
Crop Yield: 61.0
Crop Price: 70
Harvesting Month: 11
Pesticide Needed: False
Fertilizer Needed: True

4.4.2.4 Built-in Mathematical Functions

Python includes a number of built-in mathematical functions that can be used to perform basic arithmetic operations. Some of the most commonly used built-in mathematical functions in Python have been described in Table 4.2. The use of these functions has been demonstrated in Listing 4.18.

Listing 4.18	Python code demonstrating the use of built-in *abs(), max(), min(), round(), pow(), divmod(), sum()* functions in different agricultural scenarios

```
# min() function to get minimum field temperature in a week
min_field_temp = min(13.3, 20.1, 7.7, 12.5, 14.8, 11.5, 14.2)
print("Minimum Field Temperature:", min_field_temp)

# max() function to get maximum rainfall in a week
max_rainfall = max(9.2, 8.5, 10.7, 1.6, 1.0, 5.6, 3.0)
print("Maximum Rainfall:", max_rainfall)

# abs() to calculate the change in humidity in agriculture field
initial_humidity = 33.2
current_humidity = 35.7
humidity_change = abs(initial_humidity - current_humidity)
print("The change in humidity level is", humidity_change)

# round() to get air moisture level upto 2 decimal points
moisture_level = 31.6789
rounded_moisture = round(moisture_level, 2)
print("The Air Moisture Level:", rounded_moisture)

# pow() to calculate the agriculture field area
side_length = 50 # meters
area = pow(side_length, 2) # side_length^2
print(f"The area of the plot is {area} square meters.")

# divmod() to calculate total boxes and leftover apples
total_apples = 1000
box_capacity = 30
```

Table 4.2 Common Python math functions

Math function	Description
abs(number)	This function returns the absolute value of a number
max(number)	This function returns the maximum value in a collection
min(number)	This function returns the minimum value in a collection
round()	This function returns the rounded value of a number
pow()	This built-in function takes two arguments to raise the first argument number value to a specified power (i.e., the second argument number)
divmod()	The divmod() function of Python returns the quotient and remainder when dividing two numbers
sum()	The sum function returns the sum of all the elements given to this function

```
boxes, leftover = divmod(total_apples, box_capacity)
print(f"Total {boxes} full boxes & {leftover} leftover apples.")

# sum() to calculate total fertilizer use in fields
fertilizer_used = [10, 20, 15, 25, 30]
total_fertilizer = sum(fertilizer_used)
print(f"Total fertilizer used is {total_fertilizer} kg.")
```

Output of Listing 4.18:

The self-explanatory output of Listing 4.18 is shown below.

Minimum Field Temperature: 7.7
Maximum Rainfall: 10.7
The change in humidity level is 2.5
The Air Moisture Level: 31.68
The area of the plot is 2500 square meters.
Total 33 full boxes & 10 leftover apples.
Total fertilizer used is 100 kg.

4.4.2.5 Built-in String Functions

Strings represent the sequence of characters (written inside single ‘ ’ or double “ ” quotes) and are used extensively in programming. In Python, several built-in functions are available for string manipulation. Some of the most commonly used string functions of Python (shown in Table 4.3) have been explained below with the help of examples in Listing 4.19.

Table 4.3 Common Python string functions

String function	Description
len()	Returns the length of a string
count()	Returns the number of occurrences of a specific word in the string
strip()	Returns the updated string after removing leading and trailing whitespaces from a given string
capitalize()	Returns the updated string after capitalizing the first letter of a given string
lower()	Returns the updated string in all lowercase letters
upper()	Returns the updated string in all uppercase letters
replace()	Used to replace a string with another string
join()	Joins a list of strings into a single string using a specified delimiter
isalpha()	Used to check whether all the characters in the string are alphabets or not
split()	Used to split a given string into a list of substrings based on a specified delimiter
startswith()	This function returns True if a string starts with a specified substring
endswith()	This function returns True if a string ends with a specified substring

Listing 4.19	Python code demonstrates the use of common string built-in functions, which are mentioned in Table 4.3

```
# Use of len() function to check report validity
print("=============len() Function=============")
field_report = ("The pH Level of sandy soil is high.")

if(len(field_report) > 20 and len(field_report) < 80):
    print (field_report)
else:
    print("Field report should be between 20-80 characters long")

# count() function to count a word in record
print("=============count() Function=============")
pest_list = "Beetle, Aphid, Weevil, Aphid, Whitefly"
pest_attacks_count = pest_list.count("Aphid")
print("Aphid Attack Count: ", pest_attacks_count)

# Use of lower(), upper(), capitalize(), capital() functions
print("===lower(), upper(), capitalize(), capital() Function===")
plant_name = "sun FLower"
plant_name_lower = plant_name.lower()
print("Plant name in lower case:", plant_name_lower)
plant_name_capitalized = plant_name.capitalize()
print("Capitalized plant name:", plant_name_capitalized)
plant_name_title = plant_name.title()
print("Plant title:", plant_name_title)
plant_name_upper = plant_name.upper()
print("Plant name in upper case:", plant_name_upper)
```

```
# Use of replace() function
print("==========replace() Function==========")
variety_record = "Tobacco variety: KT 215 LC"
updated_variety_record = variety_record.replace("215", "209")
print("The updated crop variety:", updated_variety_record)

# Use of isalpha function
print("==========isalpha() Function==========")
crop_name = "Tobacco"
is_valid_crop_name = crop_name.isalpha()
print(is_valid_crop_name)
crop_name = "Tobacco KT 215 LC"
is_valid_crop_name = crop_name.isalpha()
print(is_valid_crop_name)

# Use of strip() function
print("============strip() Function===========")
fruits_names_list = " Apple, Banana, Guava, Cherry "
print("Fruit List:", fruits_names_list)
updated_fruits_list = fruits_names_list.strip()
print("Updated with Fruit List:",updated_fruits_list)

# Use of join() function
print("=============join() Function=============")
plant_fungal_diseases = ["Powdery mildew", "Black spot",
                                                      "Rice blast"]
disease_list = ", ".join(plant_fungal_diseases)
print("Common Plant Fungal Diseases: " + disease_list)

# Use of startwith() and endswith() functions
print("===startwith() and endswith() Function===")
variety = "Tobacco variety: KT 215 LC"
if variety.startswith("Tobacco"):
    print("This is a Tobacco variety.")
    if variety.endswith("LC"):
        print("The variety is Burley Tobacco variety")
    else:
        print("Not Burley Tobacco variety")
else:
    print("Variety is unknown")
```

Output of Listing 4.19:

The self-explanatory output of Listing 4.19 is shown below.

```
=============len() Function=============
The pH Level of sandy soil is high.
=============count() Function=============
Aphid Attack Count: 2
===lower(), upper(), capitalize(), capital() Function===
Plant name in lower case: sun flower
Capitalized plant name: Sun flower
Plant title: Sun Flower
Plant name in upper case: SUN FLOWER
==========replace() Function==========
The updated crop variety: Tobacco variety: KT 209 LC
==========isalpha() Function==========
True
False
============strip() Function===========
Fruit List: Apple, Banana, Guava, Cherry
Updated with Fruit List: Apple, Banana, Guava, Cherry
=============join() Function=============
Common Plant Fungal Diseases: Powdery mildew, Black spot, Rice blast
===startwith() and endswith() Function===
This is a Tobacco variety.
The variety is Burley Tobacco variety
```

4.5 Python Module

A Python module is a file (with .py extension) that contains constants, functions, and classes. Modules assist programmers in organizing code more effectively than functions. Therefore, instead of writing everything in a single file, programmers split the code of a large-scale project into multiple modules for better code organization, readability, reusability, namespace management (avoiding naming conflicts), and maintainability. Other than Built-in (or predefined) Python modules, Python also supports the use of User-defined (or Custom) modules.

4.5.1 User-Defined Python Modules

User-defined modules allow programmers to organize their code by defining constants, functions, and classes in a Python file (with .py extension).

4.5.1.1 Creation of Python Module

Create a Python module named *smartfarm.py* containing three functions to check whether current moisture/temperature readings are within the ideal threshold level(s) and calculate the total fertilizer amount required for a field of a given area. Considering these requirements, the code for this module is provided in Listing 4.20.

Listing 4.20	Python module ***smartfarm.py***

```
# Constants
SOIL_MOISTURE_THRESHOLD = 30 # in percentage
IDEAL_TEMPERATURE_RANGE = (20, 30) # in degrees Celsius
FERTILIZER_REQ_PER_HA = 50 # in kilograms

# Function 1: Check Soil Moisture Level
def check_soil_moisture(moisture_level):
    """
    Check if the soil moisture level is below the threshold.
    Parameters: Moisture_level:
    Current soil moisture level in percentage.
    Returns: string:
    A recommendation for irrigation based on the moisture level.
    """
    if moisture_level < SOIL_MOISTURE_THRESHOLD:
        return "Soil moisture is low. Irrigation is required."
    else:
        return "Soil moisture is adequate. No irrigation needed."

# Function 2: Analyze Temperature Suitability
def analyze_temperature(current_temp):
    """
    Check if the temperature is within the ideal range for crop
    Parameters: current_temp:
    Current temperature in degrees Celsius.
    Returns: String:
    A message indicating if the temperature is ideal or not.
    """
    if IDEAL_TEMPERATURE_RANGE[0] <= current_temp <=
                                        IDEAL_TEMPERATURE_RANGE[1]:
        return "Temperature is ideal for crop growth."
    else:
        return "Temperature is not ideal."
```

```
# Function 3: Calculate Fertilizer Requirement
def calculate_fertilizer(area_in_ha):
    """
    Calculates the total fertilizer required for a field area
    Parameters: area_in_hectares:
    Area of the field in hectares.
    Returns: float value
    Total fertilizer required in kilograms.
    """
    total_fertilizer = FERTILIZER_REQ_PER_HA * area_in_ha
    return total_fertilizer
```

4.5.1.2 Using Python Module

In the module named *smartfarm.py*, three constants and three functions have been defined. To call these functions from another Python script, it is required to import the *smartfarm.py* module. After importing *smartfarm.py* module, these three functions (with respective passing arguments, i.e., current moisture level, current temperature level, and field area in hectares) can be invoked as shown in Listing 4.21.

Listing 4.21	Python code to call functions defined in module ***smartfarm.py***

```
import smartfarm

# Example 1: Check Irrigation Needs
moisture_level = 25 # in percentage
print(smartfarm.check_soil_moisture(moisture_level))

# Example 2: Check Temperature Needs
current_temperature = 28 # in degrees Celsius
print(smartfarm.analyze_temperature(current_temperature))

# Example 3: Calculate fertilizer requirement
field_area = 10 # in hectares
fertilizer_needed = smartfarm.calculate_fertilizer(field_area)
print(f"Total fertilizer required: {fertilizer_needed} kg")
```

In Listing 4.21, the first statement is the import statement. There are different ways to use the import statement. For example,

(a) To import the entire module, you can use

```
import smartfarm
```

(b) You can also create an alias for the imported module as shown below

```
import smartfarm as sm
```

In such a case, use *sm* to call any functions defined in the ***smartfarm.py*** module as shown below

```
sm.check_soil_moisture(moisture_level)
sm.analyze_temperature(current_temperature))
sm.calculate_fertilizer(field_area)
```

(c) To import a specific function (e.g., **check_soil_moisture**) in the defined module, you can write the import statement in the following way.

```
from smartfarm.py import check_soil_moisture
```

In this case, you can only call the *check_soil_moisture function* as shown in the code snippet below.

```
from smartfarm import check_soil_moisture
moisture_level = 25 # in percentage
print(check_soil_moisture(moisture_level))
```

(d) To import all names from a module, you can write the import statement in the following way:

```
from smartfarm import *
```

However, this way is not recommended, and its use is discouraged to avoid name collisions. A name collision occurs when two different modules imported in a file contain function(s) that share the same name, as shown in Listing 4.22.

Listing 4.22	Python code demonstrating the name collision in two different modules

```
import smartfarm.py
import digitalfarm.py

# defined in smartfarm.py
FERTILIZER_PER_HECTARE = 50 # in kilograms
def calculate_fertilizer(area_in_hectares):
    total_fertilizer = FERTILIZER_PER_HECTARE * area_in_hectares
    return total_fertilizer

# defined in digitalfarm.py
FERTILIZER_NEED_PER_HA = 20 # in kilograms
def calculate_fertilizer(total_ha):
    total_fertilizer = FERTILIZER_NEED_PER_HA * total_ha
    return total_fertilizer

print(calculate_fertilizer(5))
```

In this Listing 4.22 code snippet, importing two modules containing two independent functions with the same name (*calculate_fertilizer*) leads to a collision, with the latter imported module (digitalfarm) overwriting the former (smartfarm). There are two ways to avoid such collisions, i.e., call the function with the module name or use aliases for these functions for clarity, as shown below.

(a) Call the function with the module name

```
print(smartfarm.calculate_fertilizer(5))
print(digitalfarm.calculate_fertilizer(15))
```

(b) Use the alias in the import statement for clarity

```
from smartfarm import calculate_fertilizer as smcalfr
from digitalfarm import calculate_fertilizer as digcalfr

# Calling function that is defined in smartfarm.py
print(smcalfr(5))

# Calling function that is defined in digitalfarm.py
print(digcalfr(15))
```

4.5.1.3 Use of__name__

In Python, there is a special variable called *__name__* variable (two underscores before and after the word "name") that finds its value depending on how you execute the containing script. Python allows writing a module with functions that can be used in other Python scripts, and for this reason, sometimes it is useful to know whether the script is being run directly or imported as a module. To get a clearer understanding, consider the following example. Suppose the following code is saved in a file named as *myscript.py*.

myscript.py

```
def my_function():
    print ('The value of __name__ is ' + __name__)
if __name__ == '__main__':
    my_function()
```

If you execute *myscript.py* directly, then the *if __name__ == '__main__'* conditional statement will result in *true* and the output will be '*The value of __name__ is __main__*' as a result of execution of *my_function()*. The process of this execution is shown in Fig. 4.3.

From Fig. 4.3, it becomes clear that before the execution of code written inside *myscript.py*, the *__name__* variable is set to a default value, i.e., "*main.*" As it is

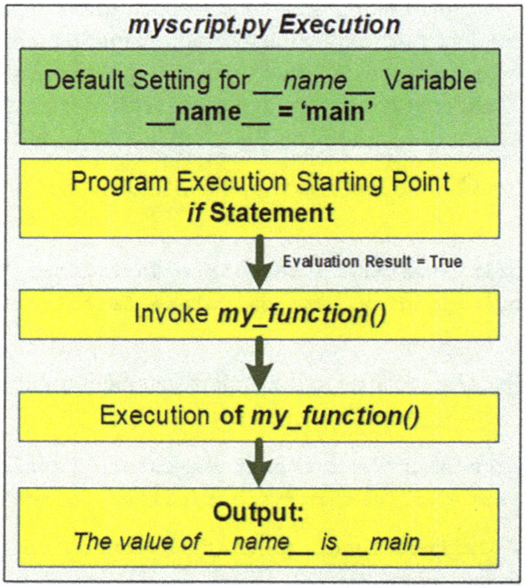

Fig. 4.3 Example demonstrating the direct execution of a Python script

already set to *main*, the condition evaluates to *true*, which ultimately results in a call to *my_function()*, and the output will be:

```
The value of __name__ is __main__
```

Now, in another scenario, if you call *my_function()* after importing this (*myscript.py*) module in another script (named as *importedscript.py*) as shown below, then the output of this function call will be:

```
The value of __name__ is myscript
```

importedscript.py

```
import myscript as ms

ms.my_function()
```

In this case, when the above code (written in *importedscript.py*) is executed, Python starts searching for a file named *myscript.py*, loads its contents, and executes the code within it. However, in *myscript.py*, the special variable *__name__* is assigned the imported module's name "*myscript*" instead of "*__main__*", because the script is being imported rather than run directly. That is why this time the condition *if __name__ == '__main__'* evaluates to false and *my_function()* will not be

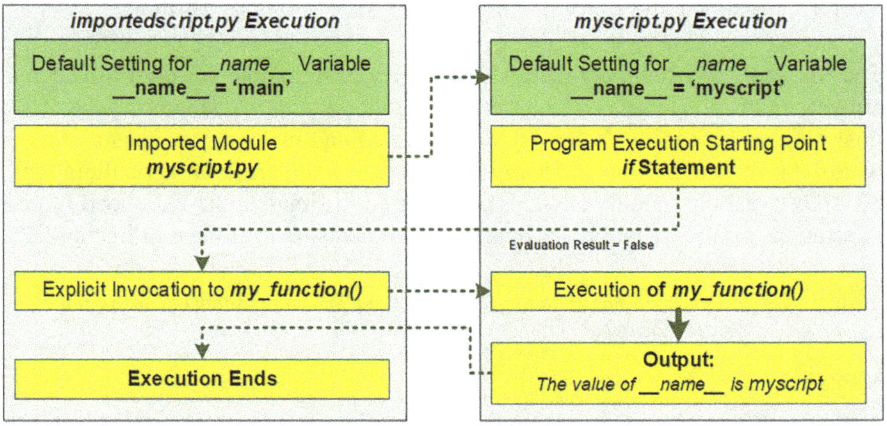

Fig. 4.4 Example demonstrating the indirect execution of a Python script

executed. However, since this function is explicitly called (as *ms.my_function()*) from the *importedscript.py*, the function executes, and the output will be:

```
The value of __name__ is myscript
```

The process illustrated in Fig. 4.4 demonstrates how Python identifies whether a script is being run directly or is being executed as part of another script where it has been imported as a module.

4.6 Built-in Modules

In Python, built-in modules are the Python files containing methods to simplify the development process by offering a diverse range of functionalities, i.e., system-level operations, file handling operations, web services development, performing mathematical calculations, and so on. Reduced development time, consistent maintainability (with regular updates), and reliability (through bug-free releases, standardization, and availability of detailed official documentation) are the key advantages of using built-in modules. The basic use of the four most commonly used Python built-in modules, i.e., *random*, *datetime*, and *math*, has been discussed below.

4.6.1 Python Random *Module*

The *random* module is used to deal with the generation of (pseudo)random numbers. This module contains several methods. The uses of a few methods of this module in the agricultural context have been shown below.

4.6.1.1 *random()* and *randint(start_number, stop_number)* Methods

The *random()* and *randint()* methods of the *random* module are used to generate random numbers. However, the difference is that the *random()* method generates a float number between *0.0* and *1.0*, and the *randint()* method generates a number within a given range from *start_number* and *stop_number*. In an agricultural context, these random module methods can be used to simulate the likelihood of pest infestations or disease outbreaks in agricultural fields, as illustrated in Listing 4.23.

Listing 4.23 Python code demonstrating the use of ***random()*** and ***randint()*** methods of the ***random*** module

```
import random

# Probability of disease outbreak in percentage
disease_outbreak_prob = random.random() * 100

# Probability of pest infestation in percentage
pest_infestation_prob = random.randint(1, 100)

print(f"Disease Outbreak Probability:{disease_outbreak_prob:.2f}.")
if disease_outbreak_prob < 30:
    print("So, Chances of disease outbreak are less.")
else:
    print("Chances of disease outbreak are high.")

print("Pest Infestation Probability:", pest_infestation_prob,".")
if pest_infestation_prob < 40:
    print("So, Chances of pest attacks are less.")
else:
    print("So, Chances of pest attacks are high.")
```

Output of Listing 4.23:

The self-explanatory output of Listing 4.23 is shown below.

Output after 1st time execution
Disease Outbreak Probability: 56.34.
Chances of disease outbreak are high.
Pest Infestation Probability: 7.
So, Chances of pest attacks are less.

Output after 2nd time execution
Disease Outbreak Probability: 33.96.
Chances of disease outbreak are high.
Pest Infestation Probability: 8.
So, Chances of pest attacks are less.

Output after 3rd time execution
Disease Outbreak Probability: 80.10.
Chances of disease outbreak are high.
Pest Infestation Probability: 44.
So, Chances of pest attacks are high.

4.6.1.2 The *seed()* Method

The *seed()* method of *random* Python module ensures the initialization of the random number generator to produce reproducible results. Within the context of agriculture, setting a seed value is used to ensure the same results generation in repeated simulations for analysis. For example, the output of the program in Listing 4.23 will be the generation of different random numbers each time you execute this code. If you want the same output each time for a particular simulation work, then you have to use a specific argument value in the *seed()* method; and for that argument value, *random()* and *randint()* methods will generate the same random numbers always, as shown in Listing 4.24.

Listing 4.24	Python code demonstrating the use of ***seed()*** method with ***random()*** and ***randint()*** methods of ***random*** module

```
import random
random.seed(55)

# Probability of disease outbreak in percentage
disease_outbreak_prob = random.random() * 100

# Probability of pest infestation in percentage
pest_infestation_prob = random.randint(1, 100)

print(f"Disease Outbreak Probability: {disease_outbreak_prob:.2f}.")
if disease_outbreak_prob < 30:
    print("So, Chances of disease outbreak are less.")
else:
    print("Chances of disease outbreak are high.")

print("Pest Infestation Probability:", pest_infestation_probab,".")
if pest_infestation_prob < 40:
    print("So, Chances of pest attacks are less.")
else:
    print("So, Chances of pest attacks are high.")
```

Output of Listing 4.24:

The self-explanatory output of Listing 4.24 is shown below.

Output after 1st time execution
Disease Outbreak Probability: 9.03.
So, Chances of disease outbreak are less.
Pest Infestation Probability: 26.
So, Chances of pest attacks are less.

Output after 2nd time execution
Disease Outbreak Probability: 9.03.
So, Chances of disease outbreak are less.
Pest Infestation Probability: 26.
So, Chances of pest attacks are less.

4.6.2 *Python* Datetime *Module*

A date in Python is not a data type of its own, but we can import a module named datetime to work with dates. Listings 4.25 and 4.26 demonstrate the use of different methods of Python's *datetime* module.

Listing 4.25 Python code demonstrating the use of ***now()*** method of ***datetime*** module

```
import datetime

current_time = datetime.datetime.now()

print("Current Time:", current_time)
print("Current Year:", current_time.year)
print("Current Month:",current_time.month)
print("Current Day:",current_time.day)
print("Current Hour:",current_time.hour)
print("Current Minute:",current_time.minute)
print("Current Second:",current_time.second)
```

Output of Listing 4.25:

After executing the code in Listing 4.25, you will get a similar output as shown below.

Current Time: 2025-03-20 21:02:15.038512
Current Year: 2025
Current Month: 3
Current Day: 20
Current Hour: 21
Current Minute: 2
Current Second: 15

Explanation
The output in this format contains year, month, day, hour, minute, second, and microsecond. Here, in the statement, the first *datetime* is the name of the module, the second *datetime* is the name of the class in this module, and *now()* is the method. The object current_time has various attributes to extract different parts of the output, i.e., year, month, date, and so on as shown in this listing.

In the datetime module, there is a method called *strftime()* that is used for formatting date objects into readable strings. A number of parameters can be passed to this method. The use of this method with various parameters is shown in Listing 4.26.

Listing 4.26 Python code demonstrating the use of ***strftime()*** method of ***datetime*** module

```
import datetime
current_time = datetime.datetime.now()

# return short version of year name
print("Year in short form:", current_time.strftime("%y"))
```

```
# return full year
print("Year is:", current_time.strftime("%Y"))

# return short version of month name
print("Month in short form :", current_time.strftime("%b"))

# return full month name
print("Month is:", current_time.strftime("%B"))

# return short name of weekday
print("Weekday name short form:",current_time.strftime("%a"))

# return full name of weekday
print("Weekday full form:", current_time.strftime("%A"))

# return number of weekday
print("Number of weekday:", current_time.strftime("%w"))

# return day of the month
print("Month day:", current_time.strftime("%d"))

# return number year's week
print("Year's week number", current_time.strftime("%V"))
```

Output of Listing 4.26:

After executing the code in Listing 4.26, you will get a similar output as shown below.

Year in short form: 25
Year is: 2025
Month in short form : Mar
Month is: March
Weekday name short form: Thu
Weekday full form: Thursday
Number of weekday: 4
Month day: 20
Year's week number 12

4.6.3 The Math *Module*

The *math* module in Python extends the list of built-in mathematical functions. To use the *math* module methods, importing the *math* module is required. The use of a few of the most common mathematical methods of Python's *math* module (shown in Table 4.4) has been described in Listing 4.27.

Table 4.4 Common math methods in the ***math*** module

Method	Description
sqrt(number)	This method returns the square root of a number
floor(number)	This method returns the round-down value of a number
ceil(number)	This method returns the round up value of a number
fabs()	This method always returns a floating-point number even if the argument is an integer. On the other hand, the abs() function returns an integer or a floating-point number depending upon the passed argument
exp()	This method returns the exponential value of a number

Listing 4.27	Python code demonstrating the use of the *math* module methods

```
import math

# Calculate diagonal of an agriculture field
field_length, field_width = 20, 40
field_diagonal = math.sqrt(field_length **2 + field_width **2)
print(f"The diagonal of agriculture field is {field_diagonal:.2f}")

# Calculate Round-up crop yield estimate
yield_estimate = 456.78
yield_rounddown_val = math.floor(yield_estimate)
print("Yield round-down value =", yield_rounddown_val)

# Calculate Round-up crop yield estimate
yield_estimate = 456.78
yield_roundup_val = math.ceil(yield_estimate)
print("Yield round-up value = ", yield_roundup_val)

# fabs() to calculate absolute error in soil pH reading
observed_pH, expected_pH = 6.5, 7.0
pH_reading_error = math.fabs(observed_pH - expected_pH)
print("The reading error is", pH_reading_error)

# exp() to predict exponential pest population increase
initial_pest_count = 100
per_day_growth_rate = 0.5
total_days = 7

'''
Exponential growth model
Pt = Initial Population (Pi) * e^(Gr * Dt)
where
Pt = Estimated pest population at time t
Pi = Initial pest population (P0)
Gr = Pest growth rate
Dt = Total number of days
'''
```

```
estimated_pest_population = initial_pest_count *
                        math.exp(per_day_growth_rate * total_days)
print(f"Estimated pest population after {total_days} days:
                              {estimated_pest_population:.2f}")
```

Output of Listing 4.27:

The self-explanatory output of Listing 4.27 is shown below.

The diagonal of agriculture field is 44.72
Yield round-down value = 456
Yield round-up value = 457
The reading error is 0.5
Estimated pest population after 7 days: 3311.55

4.7 Exercises

Problem 4.1 Define a function named *crop_info* that takes two parameters: *crop-name* and *season*. Demonstrate how arguments can be passed to this function either positionally or by using keyword arguments. The output of the program should match the structure of the following example statements:

```
"Wheat" is grown in the "Winter" season.
"Maize" is grown in the "Summer" season.
```

Problem 4.2 Define a function named *calculate_yield* that takes three parameters: *farm_id, area,* and *rate*. Besides demonstrating the use of default arguments, explain how arguments can be passed to this function positionally and by using keyword arguments. The output of the program should match the structure of the following example statements:

```
The calculated yield at Farm786 is 8 tons/hectare.
The calculated yield at Farm905 is 40 tons/hectare.
```

Problem 4.3 For plant pathologists, assessing the severity of plant diseases affecting crops is essential in determining the appropriate level of intervention. Disease severity is typically expressed as the percentage of the leaf area that is damaged. Additionally, environmental factors such as temperature and humidity can influence how rapidly a disease progresses. Based on this final percentage, plants are categorized accordingly. Develop a Python solution using functions to handle the complete process that includes the following steps:

1. Calculating the percentage of affected leaf area using the formula:

$$\text{severity\_percentage} = (\text{affected\_area} / \text{total\_area}) \times 100$$

2. Adjusting the severity score based on temperature and humidity using the formula:

$$\text{adjusted\_severity} = \text{severity} \times \text{temperature\_factor} \times \text{humidity\_factor}$$

Here, both temperature_factor and humidity_factor are set to 1.0 if the temperature is below 30 °C and the humidity is below 70%. If the temperature exceeds 30 °C, the temperature_factor increases to 1.3. Similarly, if the humidity exceeds 70%, the humidity_factor increases to 1.4.

3. Classifying the plant based on the adjusted severity score:
   (a) A plant is Healthy if the adjusted severity score is less than 10.
   (b) It is Moderately Affected if the score is between 10 and 50.
   (c) It is Severely Affected if the score is 50 or higher.
4. Recommendation for an appropriate action based on the plant's classification:
   (a) If the plant is Healthy, the recommendation message "Safe! Monitor Crop Continuously ..." should be displayed.
   (b) If Moderately Affected, the recommendation message "Apply Biological Control Techniques" should be displayed.
   (c) If Severely Affected, the recommendation message "Spray Pesticide Immediately" should be displayed.

Chapter 5
Python Data Structures

5.1 Introduction

Python data structures are also known as compound (or non-primitive) data types and are capable of holding multiple data items or values. The four common built-in Python data structures are *List*, *Tuple*, *Set*, and *Dictionary* to handle different types of data scenarios effectively. These compound data types share similarities and differences. Concerning the aspect of similarities, most of these compound data types support

- data grouping by holding multiple values, items, or data elements (of the same or different data types)
- the use of loops (*while* and *for*) over the stored elements
- various operations on stored data items for inserting, appending, storing, sorting, data manipulation, and so on
- changes in size at runtime
- membership testing through the use of *in* and *not in* keywords to check the presence or absence of a particular data item, respectively
- indexing that ultimately permits quick searching and access to a specific data item

Other than these similarities, there exist certain differences among these Python data structures, for example, not all data structures allow

- duplication of data elements
- ordering of data elements
- mutability (the ability to be modified after creation)
- indexing that assists the accessing of specific data element(s) in a sequence

The uniqueness of these data structures makes them suitable for specific tasks or use cases in Python programming. Details and relevant usage of List, Set, Tuple, and Dictionary data types in agricultural scenarios have been explained below.

M. A. Iqbal, *Python for Agriculturists*,
https://doi.org/10.1007/978-3-032-01442-9_5

5.2 List

A list in Python is a built-in data structure that supports the collection of multiple data items/elements of the same or different data types under one identifier. Python lists are useful as they offer a flexible way to store, access, and manipulate multiple data elements. Below are the key characteristics of Python lists.

- Lists can contain multiple heterogeneous data elements, i.e., integer, float, Boolean value, string, and so on.
- Duplicate data elements are allowed in lists
- Lists are ordered as the insertion of a new data element is placed at the end of the list
- Lists are dynamic and mutable as their size grows and shrinks dynamically with the possibility of adding, removing, and changing of data elements at runtime
- With zero-based indexing (the first element's index is 0), list supports both positive and negative indexing to access data elements
- Subsets of a list can be retrieved using slicing or the colon (:) operator
- Supporting the possibility of containing other lists, it is possible to create multi-dimensional lists
- Lists are iterable, which means the use of loops (*while* and *for*) is possible over the stored data elements
- Membership testing of lists is possible by determining the existence of a certain data element through the use of *in* and *not in* keywords
- Supporting comprehension, a concise way of creating new lists is possible

5.2.1 List Creation

There are various ways to create a List in Python, i.e., creating a List using square brackets [ ], using the List class constructor, using the multiplication (*) operator, and using List comprehensions.

5.2.1.1 Creating List using Square Brackets [ ]

The general syntax to create a Python list with square brackets is

```
list_name = [data_element1, data _element2, … , data_elementN]
```

Considering this syntax, below are a few examples of newly defined Python lists.

```
summer_crops = ["corn", "soybeans", "sunflowers"]
crops_and_yield = ["Wheat", 1000, "Rice", 500]
equipment_list = ["tractor", "plow", "seeder", "harvester"]
temperature_readings = [22.5, 23.0, 21.8, 22.2, 20.9]
pest_control_methods = ["Physical", "Chemical", "Biological"]
```

5.2.1.2 Creating List using List Class Constructors

The general syntax to create a Python list using the List class constructor is

```
list_name = list((data_item1, data_item2, … , data_itemN))
```

Considering this syntax, below are a few examples of newly defined Python lists.

```
winter_crops = list(("wheat", "barley", "oats"))
rainfall_levels_mm = list((5.7, 12.9, 0.8, 3.3, 9.2))
livestock_types = list(("sheep", "goats", "poultry", "pigs"))
soil_nutrients = list(("nitrogen","phosphorus","potassium"))
humidity_level_percentage = list((30, 70, 65, 90))
```

5.2.1.3 List creation with Multiplication Operator

The multiplication operator (*) can be used to create a list of repeated data elements. For example, creating a List of 10 sensors' readings with an initial 0.0% moisture level is possible in this way, as shown in Listing 5.1.

Listing 5.1	Example demonstrating the creation of a List with multiplication operator

```
# List of 10 sensors with initial 0.0% moisture level
sensors_initial_readings = [0.0] * 10
print(sensors_initial_readings)
```

Output of Listing 5.1:

[0.0, 0.0, 0.0, 0.0, 0.0, 0.0, 0.0, 0.0, 0.0, 0.0]

5.2.1.4 Creating List Using List Comprehensions

List comprehension is a smart way to create a new concise list by applying an expression to the elements of an existing iterable. The basic syntax of list comprehension is

```
list_name = [expression for item in iterable if condition
                                        == True]
```

Note: In Python, an iterable is a collection of items that you can go through one by one using a loop. String, List, Tuple, Dictionary, and so on are examples of iterables. Moreover, the if condition in the above syntax is optional.

Listing 5.2 shows different ways to create lists using list comprehensions, and Listing 5.3 provides examples of the use of list comprehensions in agricultural contexts, demonstrating how this method can be used for agriculture-related data.

Listing 5.2 | Examples demonstrating the creation of a List using list comprehension method

```
print("=====Example 1=====")
# List creation of numbers from 0 to 5.
num_list = [var for var in range(5)]
print(num_list)

print("=====Example 2=====")
# List creation containing square of values of an existing list.
values_list = [0.6, 0.9, 0.54, 0.8]
square_values_list = [val * 2 for val in values_list]
print(square_values_list)

print("=====Example 3=====")
# List creation by filtering out odd numbers from a list.
original_list = [4, 33, 9, 4, 35, 16]
odd_num_list = [val for val in original_list if val % 2 != 0]
print(odd_num_list)

print("=====Example 4=====")
# List (vector) creation from a matrix.
my_matrix = [[1, 2, 3], [4, 5, 6], [7, 8, 9]]
print(my_matrix)
my_vector = [val for row in my_matrix for val in row]
print(my_vector)
```

```
print("=====Example 5=====")
'''
List creation containing coordinate pairs representing all
possible combinations of x and y values within the range of 0 to
2. (Use of nested list comprehension)
'''
coordinates_list = [(x, y) for x in range(3) for y in range(3)]
print(coordinates_list)
```

Output of Listing 5.2:

```
=====Example 1=====
[0, 1, 2, 3, 4]
=====Example 2=====
[1.2, 1.8, 1.08, 1.6]
=====Example 3=====
[33, 9, 35]
=====Example 4=====
[[1, 2, 3], [4, 5, 6], [7, 8, 9]]
[1, 2, 3, 4, 5, 6, 7, 8, 9]
=====Example 5=====
[(0, 0), (0, 1), (0, 2), (1, 0), (1, 1), (1, 2), (2, 0), (2, 1), (2, 2)]
```

Listing 5.3	Examples demonstrating the use of the List comprehension method in agricultural contexts

```
print("===============Example 1===============")
# Filter selected crops for further processing
crops = ["Wheat", "Rice", "Corn", "Rice", "Soybean"]
suitable_crops = [crop for crop in crops if crop != "Rice"]
print(suitable_crops)

print("===============Example 2===============")
'''
An intelligent agriculture system keeps a record of the average
temperature monitored in different fields for a single day.
Moreover, the system is implemented to display a list of all the
temperature values that exceed the critical threshold (30°C). The
list comprehension method can be used to generate a list
containing temperatures that exceed the threshold value.
'''

# Recorded temperatures for 10 fields (in Celsius)
field_temperatures = [38, 32, 36, 25, 29, 31, 40]

# Critical temperature threshold
critical_temperature = 30
```

```
# List creation with values exceeding a threshold
high_temp_values = [temp_val for temp_val in field_temperatures if
                                        temp_val > critical_temperature]

# Output the results
print("Recorded Temperatures Values:", field_temperatures)
print("Temperature Readings above 30 °C:", high_temp_values)
```

Output of Listing 5.3:

```
===============Example 1===============
['Wheat', 'Corn', 'Soybean']
===============Example 2===============
Recorded Temperatures Values: [38, 32, 36, 25, 29, 31, 40]
Temperature Readings above 30 °C: [38, 32, 36, 31, 40]
```

5.2.2 Common List Operations

Table 5.1 shows common operations that can be performed on a List, and their use has been demonstrated from Listings 5.4–5.12.

5.2.2.1 Accessing List Element(s)

Each data element of the Python List has an associated number known as the index, as illustrated in the examples below.

Example 5.1 The List creation statement (to store different crop names) along with its representation in memory is shown below.

```
crop_names = ["corn", "wheat", "soybean", "rice", "cotton"]
```

index_value	0	1	2	3	4
crop_names	corn	wheat	soybean	rice	cotton

Example 5.2 The List creation statement (to store crop yields) along with its representation in memory is shown below.

```
crop_yields = [17.5, 3.8, 9.5, 13.33, 15.0]
```

index_value	0	1	2	3	4
crop_yields	17.5	3.8	9.5	13.33	15.0

Table 5.1 Common list operations

Operation	Way to implement in Python
Accessing (Individual or Multiple) List Element(s)	Using index numbers *list_name[index]* *list_name[start_index : end_index – 1]*
Traversing (all or specified) List Elements	Using ***for*** loop
Membership Testing of List Element	Using ***in*** or ***not in*** Operators
Calculating list length	Using built-in function *len(list_name)*
Finding the smallest and largest element in the list	Using built-in functions *min(list_name)* and *max(list_name)*
Calculating the sum of all elements in the list	Using built-in function *sum(list_name)*
Concatenation of lists	Using + operator
Creating multiple concatenated copies of a list	Using * operator
Comparing of two lists	Using comparison operators *(==, !=, <, <=, >, >=)*
List assignments	Using assignment (=) operator

From Examples 5.1 and 5.2, it becomes evident that the index numbers of Python Lists start from zero. This index value is very important as it is used to access data element(s) of the Python List. The general syntax of accessing individual List elements is

```
list_name[index value of list data element/item]
```

Therefore, the first element of a list (at index 0) can be accessed using the index operator [ ], such as list_name[0]. Similarly, the second element (at index 1) can be accessed using list_name[1], and so on. Moreover, it is important to keep in mind that

- the index number of the last data item of the list is always ($n - 1$) which is one less than the list's length (n).
- individual elements of a list can be accessed using negative indexing. However, in this case, the index of the last element in the List is −1 and it decreases by −1 (i.e., −2, −3, ...) as you move backward through the list.
- A colon (:) also called a slicing operator can be used with index numbers to access a specific range of data elements in a Python list. The general syntax of accessing a specific portion of a list is

```
list_name(start_index : end_index - 1 : stride)
```

In this syntax, except for one colon (:), all parameters are optional. In the case of using only one slicing operator (:) between *start_index* and *end_index – 1*, by default the *start_index* counting starts from 0 and goes up to the *end_index – 1*. The last parameter is known as the *stride* and it determines the number of data elements to skip after retrieving the first data element from the list. By default, the stride value is 1; however, you can explicitly mention any logically correct value.

Listing 5.4 provides examples to explain different ways of accessing data elements in Python.

Listing 5.4	Examples demonstrating the use of *index [ ]* operator to access List data items

```
crops_names = ["corn", "wheat", "soybean", "rice", "cotton"]

# display the first data item in the list
print("First Item:", crops_names[0])

# display the 3rd data item in the list
print("Third Item:", crops_names[3])

# display last data item in the list
print("Last Item:", crops_names[-1])

# display the 3rd last data item in the list
print("Third Last Item:", crops_names[-3])

# display list data items from index 1 to index 3
print("2nd and 3rd Items:", crops_names[1:4])

# display list data items from index 0 to index 3
print("First Four Item:", crops_names[:4])

# display list data items from index 2 to last index
print("From 3rd to Last Items:", crops_names[2:])

# display list while skipping two items each time
print("2nd and 4th Items:", crops_names[1:4:2])

# display all data items from index 0 to index 4
print("All List Items:", crops_names[:])
```

Output of Listing 5.4:

The self-explanatory output of the Listing 5.4 is shown below.

First Item: corn
Third Item: rice
Last Item: cotton
Third Last Item: soybean
2nd and 3rd Items: ['wheat', 'soybean', 'rice']
First Four Item: ['corn', 'wheat', 'soybean', 'rice']
From 3rd to Last Items: ['soybean', 'rice', 'cotton']
2nd and 4th Items: ['wheat', 'rice']
All List Items: ['corn', 'wheat', 'soybean', 'rice', 'cotton']

5.2.2.2 Traversing (All or Specified) List Elements

The *for* loop can be used to traverse all List elements or elements at particular positions in the list, one by one in each iteration. Listing 5.5 provides examples to illustrate the use of a *for* loop to access all data elements in a List.

Listing 5.5	Examples demonstrating the use of *for* loop to access List data items

```
soil_nutrients = ["Nitrogen", "Phosphorus", "Potassium"]

print("Example 1: Simple List Traversing")
# Use of for loop without built-in functions
for index in soil_nutrients:
    print("Nutrient:", index)

print("Example 2: List Traversing using Built-in Functions")
# Use of for loop with Python built-in functions (range and len)
for index in range(len(soil_nutrients)):
    print("Nutrient", index, ":", soil_nutrients[index])

print("Example 3: Traversing while Skipping Elements")
# Use of for loop to print odd-numbered position elements
for index in range(0, len(soil_nutrients), 2):
    print("Nutrient", index, ":", soil_nutrients[index])
```

Output of Listing 5.5:

The self-explanatory output of Listing 5.5 is shown below.

```
Example 1: Simple List Traversing
Nutrient: Nitrogen
Nutrient: Phosphorus
Nutrient: Potassium
Example 2: Traversing using Built-in Functions
Nutrient 0 : Nitrogen
Nutrient 1 : Phosphorus
Nutrient 2 : Potassium
Example 3: Traversing while Skipping Elements
Nutrient 0 : Nitrogen
Nutrient 2 : Potassium
```

5.2.2.3 Membership Testing of List Element

Membership testing means checking the presence of a particular data item in a given list as demonstrated in Listing 5.6.

Listing 5.6	Example demonstrating Membership Testing in list

```
# Verify if a sensor ID is in active sensor list
active_sensors = ["ID001", "ID002", "ID003"]
sensor_id = input("Enter the sensor ID to check: ").upper()

if sensor_id in active_sensors:
    print("Sensor", sensor_id, "is active.")
else:
    print("Sensor", sensor_id, "is not active.")

# Verify if a pest is in list of monitored pests
monitored_pests = ["aphid", "weevil", "armyworm"]
pest = input("Enter the pest name to check: ").lower()

if pest in monitored_pests:
    print("Yes", pest.capitalize(), "is being monitored")
else:
    print("No", pest.capitalize(), "is not being monitored.")
```

5.2.2.4 Use of len(), min(), max(), sum() Built-in Functions

Examples in Listing 5.7 are related to the use of built-in functions with Lists.

Listing 5.7	Examples demonstrating the use of built-in functions with Python List

```
# Calculate the total number of elements in a List
farm_equipment = ["tractor", "plow", "seeder", "harvester"]
print("Total equipment at farm:", len(farm_equipment))

# Calculate Min., Max., and Avg. Temperature of 5 fields
temperature_readings = [22.5, 23.0, 21.8, 22.2, 20.9]
print("Min. Temp.:", min(temperature_readings))
print("Max. Temp.:", max(temperature_readings))
readings_sum = sum(temperature_readings)
total_readings = len(temperature_readings)
print("Avg. Temp.:", round(readings_sum/total_readings, 2))
```

Output of Listing 5.7:

The self-explanatory output of Listing 5.7 is shown below.

Total equipment at farm: 4
Min. Temp.: 20.9
Max. Temp.: 23.0
Avg. Temp.: 22.08

5.2.2.5 Use of + and * Operators with Python Lists

The arithmetic operator + can be used to concatenate two Lists and the arithmetic operator * is used to concatenate different copies of Lists as shown in Listing 5.8.

Listing 5.8	Example demonstrating the use of + and * operators with List

```
summer_crops = ["corn", "soybeans", "sunflowers"]
winter_crops = ["wheat", "oat", "barley"]

# Use of + to concatenate lists of summer and winter crops
yearly_crops = summer_crops + winter_crops
print("Yearly Crops:", yearly_crops)

# Use of * to create copies of crop rotation plan
crop_rotation_plan = ["Wheat", "Barley"]

# create 3 concatenated copies of crop rotation
three_year_rotation_plan = 3 * crop_rotation_plan
print("Three Year Crop Rotation Plan:", three_year_rotation_plan)
```

Output of Listing 5.8:

The self-explanatory output of Listing 5.8 is shown below.

Yearly Crops: ['corn', 'soybeans', 'sunflowers', 'wheat', 'oat', 'barley']
Three Year Crop Rotation Plan: ['Wheat', 'Barley', 'Wheat', 'Barley', 'Wheat', 'Barley']

5.2.2.6 Use of Comparison Operator with Python List

Comparison operators (i.e., ==, !=, >, >=, <, <=) can be used to compare the elements of lists. In such scenarios, the comparison follows lexicographical (dictionary) ordering. For example, the comparison starts with the first elements of both lists; if the data elements are different, the comparison result is based on these elements. However, the next elements are compared if the first elements are the same, and this process continues until a difference is found or one of the lists runs out of elements. Examples in Listing 5.9 are related to the demonstration of the use of comparison operators with Lists.

Listing 5.9	Examples demonstrating membership testing in list

```
# Example 1: Compare nutrient levels in two soil samples
print("Example1: Nutrient's Level Comparison")

# N, P, K levels in soil sample 1 (in mg/kg)
sample1 = [30, 40, 50]

# N, P, K levels in soil sample 2 (in mg/kg)
sample2 = [35, 40, 45]

if sample1 == sample2:
    print("Soil samples have same nutrient composition.")
else:
    print("Soil samples have different nutrient composition.")

# Example 2: Comparing sensors' readings
print("Example2: Sensors' moisture reading comparison")

# Sensor 1 readings for Soil Moisture (in %age)
sensor1_readings = [45, 50, 55]

# Sensor 2 readings for Soil Moisture (in %age)
sensor2_readings = [40, 48, 52]

if sensor1_readings > sensor2_readings:
    print("Sensor 1 readings are lower than Sensor 2.")
else:
    print("Sensor 2 readings are same or higher than Sensor 1.")

# Example 3: Compare crop yield estimates for two years
print("Example3: Comparison of yield estimate for two years")

# Maize crop yield (kg/acre) in Summer & Winter of 2023
year2023_yield = [1000, 1200]

# Maize crop yield (kg/acre) in Summer & Winter of 2024
year2024_yield = [950, 1250]

if year2023_yield >= year2024_yield:
    print("Year 2023 has equal/better yield projections.")
else:
    print("Year 2024 has better yield projections.")

# Example4: Comparing pesticide usage plan
print("Example4: Comparison of two plans for pesticide usage")

# Pesticide usage plans (in liters) for three fields
plan1 = [2, 2, 4]
plan2 = [2, 2, 4]
```

```
if plan1 >= plan2:
    print("Plan1 uses equal/more pesticides than Plan2.")
else:
    print("Plan2 uses more pesticides compared to Plan1.")
```

Output of Listing 5.9:

The self-explanatory output of Listing 5.9 is shown below. (Note: It might be possible that the comparison result(s) in the following output is not logically correct because all comparisons are based on lexicographical ordering.)

```
Example1: Nutrient's Level Comparison
Soil samples have different nutrient composition.
Example2: Sensors' moisture reading comparison
Sensor 1 readings are lower than Sensor 2.
Example3: Comparison of yield estimate for two years
Year 2023 has equal/better yield projections.
Example4: Comparison of Two Plans for Pesticide Usage
Plan1 uses equal/more pesticides than Plan2.
```

5.2.2.7 Use of Assignment Operator with Python List

The assignment operator (=) in Python can be used in two ways: (1) to modify the List element and (2) to assign a list reference to another list, as explained in Listing 5.10 and Listing 5.11, respectively.

Listing 5.10	Example demonstrating the modification of Python List elements

```
# Example of using = operator to update a List's content
livestock_animals = ["cattle", "sheep", "goat", "poultry", "pig"]
print("List of Livestock Animals:", livestock_animals)

# Code to replace "cattle" with "cow"
for index in range(len(livestock_animals)):
    if(livestock_animals[index] == "cattle"):
        livestock_animals[index] = "cow"
    else:
        continue

print("Updated List of Livestock Animals:", livestock_animals)
```

Output of Listing 5.10:

The self-explanatory output of Listing 5.10 is shown below.

```
List of Livestock Animals: ['cattle', 'sheep', 'goat', 'poultry', 'pig']
Updated List of Livestock Animals: ['cow', 'sheep', 'goat', 'poultry', 'pig']
```

Listing 5.11	Example demonstrating the use of the assignment (=) operator to copy one List's reference to another List

```
livestock_animals = ["cattle", "sheep", "goat", "poultry", "pig"]
print("List of Livestock Animals:", livestock_animals)

# Defining a new List
domestic_animals = ["None"]

# Assigning livestock_animal List to the domestic_animal List
domestic_animals = livestock_animals
print("List of Domestic Animals:", domestic_animals)

# Replacing "cattle" with "cow" in livestock_animals List
for index in range(len(livestock_animals)):
    if(livestock_animals[index] == "cattle"):
        livestock_animals[index] = "cow"
    else:
        continue

# Adding 'cat' and 'horse' in domestic_animals List
domestic_animals.append("cat")
domestic_animals.append("horse")

# Display of domestic_animals and listock_animal lists
print("Updated List of Livestock Animals:", livestock_animals)
print("Updated List of Domestic Animals:", domestic_animals)
```

Output of Listing 5.11:

List of Livestock Animals: ['cattle', 'sheep', 'goat', 'poultry', 'pig']
List of Domestic Animals: ['cattle', 'sheep', 'goat', 'poultry', 'pig']
Updated List of Livestock Animals: ['cow', 'sheep', 'goat', 'poultry', 'pig', 'cat', 'horse']
Updated List of Domestic Animals: ['cow', 'sheep', 'goat', 'poultry', 'pig', 'cat', 'horse']

Explanation
From the output, it is clear that using the = operator to assign one list to another does not create a steparate copy of the original list. Instead, it assigns the reference of the original list to the new list name. As a result, both references (list names) point to the same list in memory (as illustrated in Fig. 5.1). This is known as a Shallow Copy, meaning any modifications made to one list are also reflected in the other.

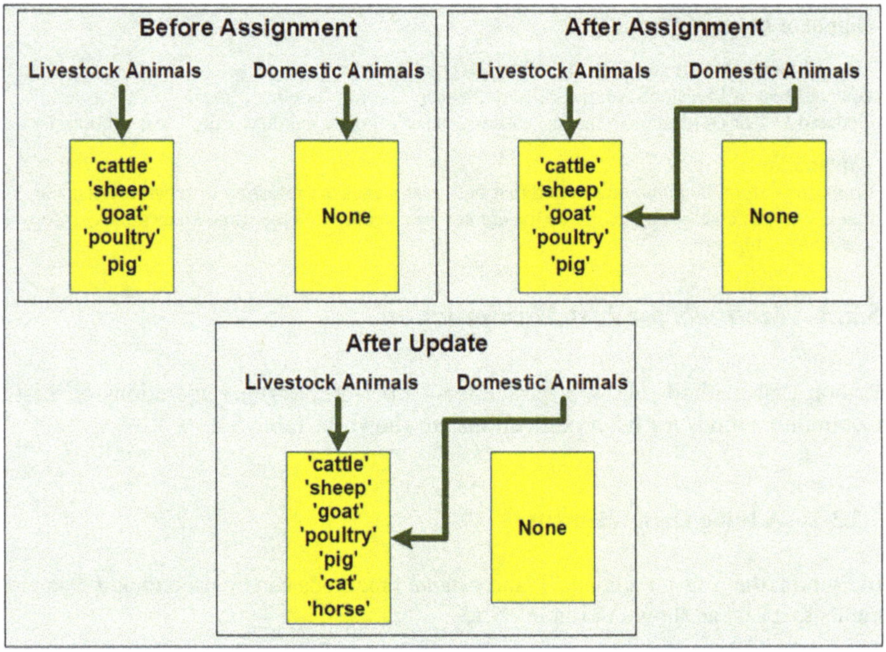

Fig. 5.1 Shallow Copy representation

There are multiple ways to cope with the issue of shallow copying (shown in Listing 5.11). One of the solutions is to use the *for* loop to copy all elements of one List to another (known as deep copying), as shown in Listing 5.12.

Listing 5.12	Example demonstrating the use of *for* loop to copy one List's content (data items) to another list

```
livestock_animals = ["cattle", "sheep", "goat", "poultry", "pig"]
print("List of Livestock Animals:", livestock_animals)

# Defining a new List
domestic_animals = []

for index in range(len(livestock_animals)):
    domestic_animals.append(livestock_animals[index])

# Replacing "cattle" with "cow" in livestock_animals List
livestock_animals[0] = "cow"

# Adding 'cat' and 'horse' in domestic_animals List
domestic_animals.append("cat")
domestic_animals.append("horse")

# Display of domestic_animals and listock_animal lists
print("Updated List of Livestock Animals:", livestock_animals)
print("Updated List of Domestic Animals:", domestic_animals)
```

Output of Listing 5.12:

List of Livestock Animals: ['cattle', 'sheep', 'goat', 'poultry', 'pig']
Updated List of Livestock Animals: ['cow', 'sheep', 'goat', 'poultry', 'pig']
Updated List of Domestic Animals: ['cattle', 'sheep', 'goat', 'poultry', 'pig', 'cat', 'horse']

Explanation
From this output, it can be interpreted that both List names are referring to separate lists, and that is why the changes made in one list are not reflected in the other (known as the DeepCopy), as shown in Fig. 5.2.

5.2.3 *Methods for List Manipulation*

Python List methods allow programmers to perform various operations on Lists. Common methods for list modifications are shown in Table 5.2.

5.2.3.1 Adding Data Elements

In Python, the *insert()*, *append()*, and *extend()* methods can be used to add data elements to a List as shown in Listing 5.13.

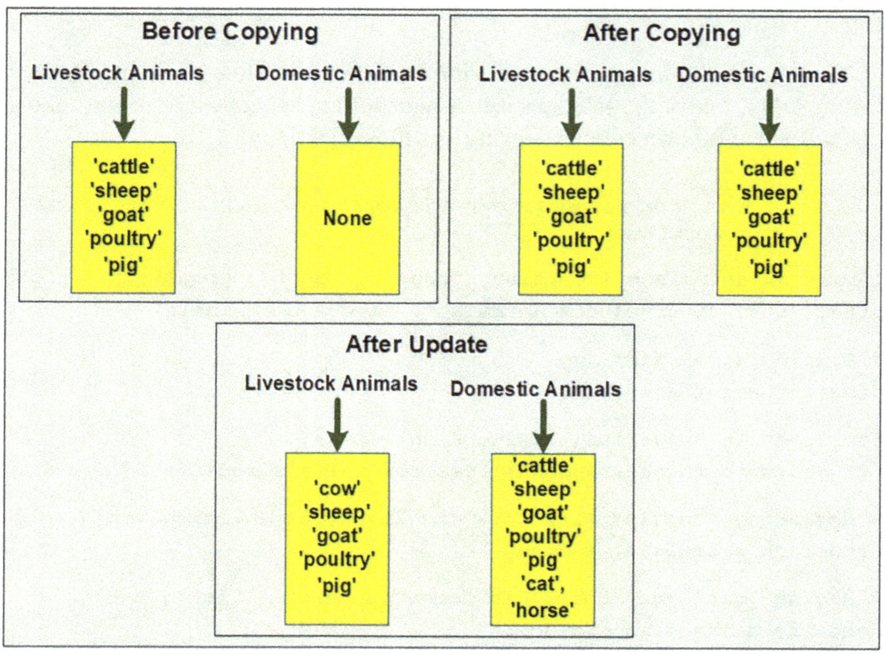

Fig. 5.2 Deep Copy representation

Table 5.2 Common list methods

Action	Method	Description
Adding elements	append (X)	Add a new data element with the value *X* at the end of the list
	insert (index, X)	Add a new data element value *X* at a specific index
	extend (iterable)	Add all data elements from an iterable to the end of the list
Removing elements	remove (X)	Removes the first occurrence of the data element of value X from the list
	pop([index])	Removes and returns the data element with index *i*. By default (in the absence of index), it removes and returns the last item from the list
	clear()	Removes all the data items from the list
List manipulation and processing methods	reverse()	Reverses the order of all the data elements in the list
	count(value)	Returns the number of occurrences of data elements with value *X* in the list.
	sort()	Used to sort the elements of the list
	index (X, [start], stop])	Returns the first index where the data element of value *X* is found in the list. Optional start and stop arguments of this method can be used to indicate a subrange of the list to search that element
	copy()	Returns a copy of the list

Listing 5.13 Example demonstrating the addition of data elements to Python List elements

```
print("================insert(index, value) Method=============")
# Example of using insert(index, value)
'''
Script for an agriculturist who wants to modify the crop rotation
list by inserting two crops into a list (at the 2nd and 3rd
index) to record the actual rotation plan.
'''
crop_rotation = ["Wheat", "Barley", "Maize"]
print("Original Crop Rotation List: ", crop_rotation)
# Adds "millet" at index 2
crop_rotation.insert(2, "millet")
# Adds "millet" at index 3
crop_rotation.insert(3, "sorghum")

print("Modified Crop Rotation List: ", crop_rotation)

print("================append(value) Method================")
# Example of using append(value)
'''
```

```
Script for a soil scientist who wants to modify the moisture
reading list by adding two more readings to a list (at the end of
the list).
'''
# List of soil moisture readings (in percentage)
moisture_readings = [35, 40, 38]
new_reading = 0

# Display the original moisture_reading list
print("Original Moisture Readings List:", moisture_readings)

# For user to input three moisture readings
for index in range(3):
    new_reading = eval(input("Enter new moisture reading: "))
    moisture_readings.append(new_reading)

# Display the modified moisture_reading list
print("Modified Moisture Readings List:", moisture_readings)

# Display moisture_reading list elements one by one.
for index in range(len(moisture_readings)):

    print("Moisture_readings", moisture_readings[index])

print("================extend(iterable) Method================")
'''
Example of using extend(iterable)
Script for ranchers who want to consolidate animal food data of
two barns
'''
# Feed consumption data (in kg) for two barns
feed_data = []
barn1_feed = [100, 110, 90]
barn2_feed = [190, 130, 162]

# Extend feed_data list with both other barn lists
feed_data.extend(barn1_feed)
feed_data.extend(barn2_feed)

print("Consolidated Feed Data:", feed_data)
print("==============================================")
```

Output of Listing 5.13:

The self-explanatory output of Listing 5.13 is shown below.

```
================insert(index, value) Method==================
Original Crop Rotation List: ['Wheat', 'Barley', 'Maize']
Modified Crop Rotation List: ['Wheat', 'Barley', 'millet', 'sorghum', 'Maize']
================append(value) Method======================
Original Moisture Readings List: [35, 40, 38]
Enter new moisture reading:29
Enter new moisture reading:31
Enter new moisture reading:33
Modified Moisture Readings List: [35, 40, 38, 29, 31, 33]
Moisture_readings 35
Moisture_readings 40
Moisture_readings 38
Moisture_readings 29
Moisture_readings 31
Moisture_readings 33
================extend(iterable) Method================
Barn 1 Feed Data: [100, 110, 90, 190, 130, 162]
Barn 2 Feed Data: [100, 110, 90, 190, 130, 162]
Consolidated Feed Data: [100, 110, 90, 190, 130, 162]
=================================================
```

5.2.3.2 Removing Data Elements

The use of the *remove()*, *pop()*, and *clear()* methods of the Python List has been illustrated in Listing 5.14.

Listing 5.14	Example demonstrating the removal of data elements to Python List elements

```
print("================remove(string) Method================")
# Example of using remove(string) method
'''
Script for a rancher who wants to remove a particular livestock
from the list
'''
livestock_types = ["cattle", "sheep", "goats", "poultry", "pigs"]
print("Original Livestock Types List:", livestock_types)
livestock_types.remove("cattle")
livestock_types.remove("pigs")
print("Modified Livestock Types List:", livestock_types)
print("================pop(index) Method================")
# Example of using pop(index) method
'''
Script for a gardener who wants to remove a vegetable stored at a
particular index
```

```
'''
summer_crops = ["corn", "soybeans", "sunflowers", "carrots"]
print("Original List of Summer Crops:", summer_crops)
crop = summer_crops.pop(3)
print("Modified List of Summer Crops:", summer_crops)

print("================clear() Method================")

# Example of using clear() method
'''
Script for a gardener who wants to remove all items of
vegetable list
'''
vegetables_list = ["cauliflower", "cucumber", "tomato", "potato"]
print("Original List of Vegetables:", vegetables_list)
vegetables_list.clear()
print("Modified List of Vegetables:", vegetables_list)
```

Output of Listing 5.14:

The self-explanatory output of Listing 5.14 is shown below.

```
================remove(string) Method================
Original Livestock Types List: ['cattle', 'sheep', 'goats', 'poultry', 'pigs']
Modified Livestock Types List: ['sheep', 'goats', 'poultry']
================pop(index) Method================
Original List of Summer Crops: ['corn', 'soybeans', 'sunflowers', 'carrots']
Modified List of Summer Crops: ['corn', 'soybeans', 'sunflowers']
================clear() Method================
Original List of Vegetables: ['cauliflower', 'cucumber', 'tomato', 'potato']
Modified List of Vegetables: []
```

5.2.3.3 List Processing Methods

The use of the *reverse(), count(), sort(), index(), copy()* methods of Python List has been illustrated in Listing 5.15.

Listing 5.15	Example demonstrating the use of Python List processing methods

```
fruits_list = ["lime","apple","Fig","pear","plum","apple"]

print("Original Fruits List: ", fruits_list)

print("================count() Method================")
'''
Script for a farmer to calculate the total number of a particular
fruit in the List
'''
# Using count(list) method to count apples in list
print("Total apples in the list: ", fruits_list.count("apple"))
```

```
print("================index() Method================")
'''
Script for a farmer to search the presence of a particular fruit
in the List
'''
# Searching position of a fruit in list
print("Searching apple in the whole list")
print("Apple is at position ", fruits_list.index("apple"))

# Searching a fruit in a sub-range of list
print("Searching apple in the sub-range of list")
print("Apple is at position", fruits_list.index("apple",2,6))

print("================sort() Method================")
# Sorting the fruits list in ascending/descending order
# Example of sorting the list in dictionary order
print("Sorting fruit list in ascending oder")
fruits_list.sort()
print("Sorted List (in ascending order): ", fruits_list)

# Sort the list in reverse dictionary order
print("Sorting fruit list in descending oder")
fruits_list.sort(reverse=True)
print("Sorted List (in descending order): ", fruits_list)

print("================copy() Method================")
# Copying the fruits list into another list

# Example of copying the list entries into another list
print("Copying the fruit list")
my_fruits = fruits_list.copy()
print("Copied Fruit List ", my_fruits)
```

Output of Listing 5.15:

The self-explanatory output of Listing 5.15 is shown below.

```
Original Fruits List: ['lime','apple', 'fig','pear','plum','apple']
================count() Method================
Total apples in the list: 2
================index() Method================
Searching apple in the whole list
Apple is at position 1
Searching apple in the sub-range of list
Apple is at position 5
================sort() Method================
Sorting fruit list in ascending oder
Sorted List (in ascending order): ['apple', 'apple', 'fig', 'lime', 'pear', 'plum']
Sorting fruit list in descending oder
Sorted List (in descending order): ['plum', 'pear', 'lime', 'fig', 'apple', 'apple']
================copy() Method================
Copying the fruit list
Copied Fruit List ['plum', 'pear', 'lime', 'fig', 'apple', 'apple']
```

5.2.4 *Two-Dimensional Lists*

In Python, a multidimensional list is a type of data structure where each element can itself be a list, often referred to as a "list of lists." This structure is particularly helpful for organizing and displaying information in rows and columns (of tables or matrices). Dimension flexibility and hierarchical organization of storing data/information are the main characteristics of a multidimensional list.

5.2.4.1 Creating and Accessing a Two-Dimensional (2D) List

Similar to the one-dimensional list, there are various ways to create a 2D list in Python, i.e., using square brackets [ ], using the List class constructor, using the multiplication (*) operator, and using List comprehensions.

(a) Using Square Brackets [ ]

The general syntax to create a Python 2D list with square brackets is given as follows:

```
list_name = [
    [row1_value1, row1_value2, … , row1_valueN], # Row 1
    [row2_value1, row2_value2, … , row2_valueN], # Row 2
    ...
    [rowN_value1, rowN_value2, … , rowN_valueN], # Row N
]
```

(b) Using list() Constructor

The list() constructor can be used to convert an iterable into a list, allowing us to create a 2D list dynamically.

The general syntax to create a Python 2D list using list constructor is given as follows:

```
list_name = list([
    [row1_value1, row1_value2, … , row1_valueN], # Row 1
    [row2_value1, row2_value2, … , row2_valueN], # Row 2
    ...
    [rowN_value1, rowN_value2, … , rowN_valueN], # Row N
])
```

(c) Using List Comprehension

A concise way to create a 2D list is through List comprehension. The general syntax of creating a list using List comprehension is:

```
List_name = [expression for item in iterable if condition]
```

Considering these syntaxes, a few examples of defining 2D lists are shown in Listing 5.16.

Listing 5.16 | Examples demonstrating the creation of 2D Python Lists

```
print("==========================================================")
print("Examples of creating 2D List using square brackets []")
print("==========================================================")
'''
Example 1: A two-dimensional list to store different nutrient
(N, P, and K) levels in 3 agricultural fields.
'''

soil_nutrient_levels = [
    [6.5, 7.0, 6.8], # Field 1 nutrient levels
    [6.2, 6.7, 7.1], # Field 2 nutrient levels
    [6.4, 6.9, 7.0] # Field 3 nutrient levels
]
print(soil_nutrient_levels)
'''
Example 2: A two-dimensional list to store yield of 3 regions
on a farm where each region contains 3 crops (Rice, Maize,
and Wheat).
'''
farm_yield = [
    [3.5, 4.2, 3.8], # Region 1: Rice, Maize, Wheat
    [2.9, 3.7, 4.1], # Region 2: Rice, Maize, Wheat
    [3.2, 4.0, 3.9] # Region 3: Rice, Maize, Wheat
]
print(farm_yield)
print("==========================================================")
print("Examples of creating 2D List using List constructor")
print("==========================================================")
'''
Example 1: A two-dimensional list to store 3 levels of Nitrogen
(N), Phosphorus (P), Potassium (K) on 3 different
agriculture farms.
'''
soil_nutrient_levels = list([
    list(["Low", "Medium", "High"]), # Field 1: N, P, K
    list(["Medium", "High", "Low"]), # Field 2: N, P, K
    list(["High", "Low", "Medium"]) # Field 3: N, P, K
])
print(soil_nutrient_levels)
```

```
'''
Example 2: Creating a two-dimensional list (initialized with 0)
to later store the pesticide cost for 3 different fields.
'''
rows, cols = 3, 3
pesticide_cost = list(list(0 for _ in range(cols)) for _ in
                                                        range(rows))
print(pesticide_cost)

'''
Example 3: Creating a two-dimensional list (initialized with 0)
to later store the crop yields for 3 different fields.
'''
crop_yields = list([["Field " + str(i+1), [round(j * 0, 1) for j
                                in range(3)]] for i in range(3)])
print(crop_yields)

print("==========================================================")
print("Examples of creating 2D List using List comprehensions")
print("==========================================================")
'''
Example 1: Creating a two-dimensional list (initialized with 0)
to later store fertilizer cost of 3 months at 3 farms.
'''
rows, cols = 3, 3
fertilizer_cost = [[0 for _ in range(cols)] for _ in range(rows)]
print(fertilizer_cost)
'''
Example 2: Creating a two-dimensional list (initialized with 0)
to later store the rainfall data for a week at 3 farms
'''
rainfall_data = [[round(i * 0, 1) for i in range(7)] for farm in
                                                          range(3)]
print(rainfall_data)
```

Output of Listing 5.16:

```
==================================================
Examples of creating 2D List using square brackets []
==================================================
[[6.5, 7.0, 6.8], [6.2, 6.7, 7.1], [6.4, 6.9, 7.0]]
[[3.5, 4.2, 3.8], [2.9, 3.7, 4.1], [3.2, 4.0, 3.9]]
==================================================
Examples of creating 2D List using List constructor
==================================================
[['Low', 'Medium', 'High'], ['Medium', 'High', 'Low'], ['High', 'Low', 'Medium']]
[[0, 0, 0], [0, 0, 0], [0, 0, 0]]
[['Field 1', [0, 0, 0]], ['Field 2', [0, 0, 0]], ['Field 3', [0, 0, 0]]]
====================================================
Examples of creating 2D List using List comprehensions
====================================================
[[0, 0, 0], [0, 0, 0], [0, 0, 0]]
[[0, 0, 0, 0, 0, 0, 0], [0, 0, 0, 0, 0, 0, 0], [0, 0, 0, 0, 0, 0, 0]]
```

5.2.4.2 Accessing 2D List Element(s)

Each data element of Python's 2D List has two associated numbers known as the index, as illustrated in the example below.

Example 5.3 The 2D List representation in Python and memory (to store irrigation water applied (in mm) for 3 fields over 4 weeks) is shown below.

2D List Representation in Python

```
irrigation_water = [
    [30, 40, 35, 38], # Field1: Water (mm) for 4 weeks
    [25, 38, 32, 36], # Field2: Water (mm) for 4 weeks
    [28, 36, 30, 34] # Field3: Water (mm) for 4 weeks
]
```

2D List Representation in Memory

Column index / Row index	Week1 (index [0])	Week2 (index [1])	Week3 (index [2])	Week4 (index [3])
Field1 (index [0])	30	40	35	38
Field2 (index [1])	25	38	32	36
Field3 (index [2])	28	36	30	34

From this 2D list representation in memory, it becomes clear that the (row and column) index numbers of Python 2D-Lists start from zero. These index values are very important as these are used to access the Python List's data element(s). The general syntax of accessing individual 2D List element(s) is

```
list_name[row_index of list data item][column_index of list
                        data item]
```

Listing 5.17 provides examples to explain different ways of accessing data element(s) in a Python program.

Listing 5.17	Examples demonstrating the use of the row and column index numbers to access List data items

```
irrigation_data = [
    [30, 40, 35, 38], # Field1:Water (mm) for 4 weeks
    [25, 38, 32, 36], # Field2:Water (mm) for 4 weeks
    [28, 36, 30, 34] # Field3:Water (mm) for 4 weeks
]

print("Field1: Irrigation during Week 3:", irrigation_data[0][2])

print("Field2: Irrigation during Week 2:", irrigation_data[2][1])

# Slicing irrigation data for first 3 weeks of Field 1
print("Field1: Irrigation for first 3 weeks:", irrigation_
                                                  data[0][:3])

# Slicing irrigation data for all fields in last 2 weeks
print("Irrigation for last 2 weeks across all fields:", [row[-2:]
                                    for row in irrigation_data])

# Slicing irrigation data for first two fields (all weeks)
print("Field 1 and Field 2 Irrigation:", irrigation_data[:2])
```

Output of Listing 5.17:

The self-explanatory output of Listing 5.17 is shown below.

Field1: Irrigation during Week 3: 35
Field2: Irrigation during Week 2: 36
Field1: Irrigation for first 3 weeks: [30, 40, 35]
Irrigation for last 2 weeks across all fields: [[35, 38], [32, 36], [30, 34]]
Field 1 and Field 2 Irrigation: [[30, 40, 35, 38], [25, 38, 32, 36]]

5.2.4.3 Traversing 2D List Elements

The nested *for* loop can be used to traverse all List elements one by one in each iteration. Listing 5.18 provides examples to illustrate the access of data elements in a 2D List.

Listing 5.18	Examples demonstrating the traversal of 2D List with and without a nested *for* loop

```
# Display irrigation data using a nested for loop
print("=====================================================")
print("Display irrigation data using nested for loop")
print("=====================================================")
for row in range(len(irrigation_data)) :
    for col in range(len(irrigation_data[row])) :
        print(irrigation_data[row][col], end="\t\t")
    print()

# Display irrigation data without using a nested for loop
print("===================================================")
print("Display whole irrigation data without for loop")
print("===================================================")

for record in irrigation_data:
    print(record)
```

Output of Listing 5.18:

The self-explanatory output of Listing 5.18 is shown below.

```
=================================================
Display irrigation data using nested for loop
=================================================
30      40          35          38
25      38          32          36
28      36          30          34
================================================
Display whole irrigation data without using nested for loop
================================================
[30, 40, 35, 38]
[25, 38, 32, 36]
[28, 36, 30, 34]
```

5.2.4.4 Membership Testing of 2D List Element

Membership testing means checking the presence of a particular data item in a given list, as demonstrated in Listing 5.19.

Listing 5.19	Examples demonstrating the Membership Traversal in 2D List

```
irrigation_data = [
    [30, 40, 35, 38], # Field1 water (mm) for 4 weeks
    [25, 38, 32, 36], # Field2 water (mm) for 4 weeks
    [28, 36, 30, 34] # Field3 water (mm) for 4 weeks
]

# To check if a whole sublist exists in a 2D List
print("=============================================")
print("Checking if a whole sublist exists in a 2D List")
print("=============================================")
print([38, 32, 36] in irrigation_data)
print([28, 36, 30] in irrigation_data)
print([28, 36, 30, 34] in irrigation_data)

# Row-wise searching for a specific data item in 2D List
print("=====================================================")
print("Checking if a specific data item exists in a 2D List")
print("=====================================================")
value = int(input("Enter value to search in 2D List: "))
is_present = any(value in row for row in irrigation_data)
print(is_present)

# To check if a value exists in a flattened 2D List
print("=====================================================")
print("Checking specific item in flattened 2D List")
print("=====================================================")
from itertools import chain

# Flattening and checking membership
flattened_list = chain.from_iterable(irrigation_data)
print(value in flattened_list)
print(value in flattened_list)
```

Output of Listing 5.19:

The self-explanatory output of Listing 5.19 is shown below.

```
============================================
Checking if a whole sublist exists in a 2D List
============================================
False
False
True
==================================================
Checking if a specific data item exists in a 2D List
==================================================
Enter value to search in 2D List: 32
True
========================================================
Checking specific item in flattened 2D List
========================================================
True
False
```

5.2.4.5 Common Operations on 2D List

Python List methods allow programmers to perform various operations on 2D Lists. The use of common methods for list modifications is shown in Listing 5.20.

Listing 5.20	Examples demonstrating how to perform common operations on 2D list

```
'''
A 2-D list containing soil nutrient levels information of three
agricultural fields
'''
soil_nutrient_levels = [
    [6.5, 7.0, 6.8], # Field 1 nutrient (N, P, K) levels
    [6.2, 6.7, 7.1], # Field 2 nutrient (N, P, K) levels
    [6.4, 6.9, 7.0] # Field 3 nutrient (N, P, K) levels
]

# Display soil nutrient level data in tabular form
print("====================================")
print("Soil Nutrient Level Data of 3 Fields")
print("====================================")
for record in soil_nutrient_levels:
    print(record)

# Add new data row i.e., Field 4 nutrient levels
soil_nutrient_levels.append([5.0, 5.5, 7.5])
print("====================================")
print("Soil Nutrient Level Data of 4 Fields")
```

```
print("======================================")
for record in soil_nutrient_levels:
    print(record)

# Reverse a specific sublist i.e., row 4 of the list
soil_nutrient_levels[3].reverse()
print("====================================================")
print("Soil Nutrient Level Data of 4 Fields in Reverse Order")
print("====================================================")
for record in soil_nutrient_levels:
    print(record)

'''
Change value of a specific data value For example, change value of
row 1 and col 2 data element from 7.0 to 9.9
'''
print("====================================================")
print("Soil Nutrient Level Data after changing a specific value")
print("====================================================")
soil_nutrient_levels[0][1] = 9.9

for record in soil_nutrient_levels:
    print(record)
print("=============List Sorting Examples=============")

print("==============================")
print("Sort entire 2D List row-by-row")
print("==============================")

rowise_sorted_list = [sorted(row) for row in soil_
nutrient_levels]
for record in rowise_sorted_list:
    print(record)

print("====================================")
print("Sort entire 2D List column-by-column")
print("====================================")

for row in soil_nutrient_levels:
    row.sort() #using sort() method of 1D List
soil_nutrient_levels.sort()
for record in soil_nutrient_levels:
    print(record)
```

Output of Listing 5.20:

The self-explanatory output of Listing 5.20 is shown below.

```
===================================
Soil Nutrient Level Data of 3 Fields
===================================
[6.5, 7.0, 6.8]
[6.2, 6.7, 7.1]
[6.4, 6.9, 7.0]
===================================
Soil Nutrient Level Data of 4 Fields
===================================
[6.5, 7.0, 6.8]
[6.2, 6.7, 7.1]
[6.4, 6.9, 7.0]
[5.0, 5.5, 7.5]
=================================================
Soil Nutrient Level Data of 4 Fields in Reverse Order
=================================================
[6.5, 7.0, 6.8]
[6.2, 6.7, 7.1]
[6.4, 6.9, 7.0]
[7.5, 5.5, 5.0]
=================================================
Soil Nutrient Level Data after changing a specific value
=================================================
[6.5, 9.9, 6.8]
[6.2, 6.7, 7.1]
[6.4, 6.9, 7.0]
[7.5, 5.5, 5.0]
============List Sorting Examples============
=============================
Sort entire 2D List row-by-row
=============================
[6.5, 6.8, 9.9]
[6.2, 6.7, 7.1]
[6.4, 6.9, 7.0]
[5.0, 5.5, 7.5]
=================================
Sort entire 2D List column-by-column
=================================
[5.0, 5.5, 7.5]
[6.2, 6.7, 7.1]
[6.4, 6.9, 7.0]
[6.5, 6.8, 9.9]
```

5.3 Tuple

Like a Python List, a Tuple is a Python data structure that is used to store multiple homogeneous/heterogeneous data elements that are ordered and can be duplicated, but unlike Python lists, data elements in a Tuple are unchangeable (fixed or immutable). Due to this reason, the length of Python Tuples is also fixed, and addition, deletion, replacement, reordering, and so on of data elements is not possible. Nevertheless, for some context, there is a need to store data items in a Tuple to avoid accidental addition, deletion, and replacement of data elements.

5.3.1 *Tuple Creation*

The general syntax of creating a Python Tuple is shown as follows:

```
identifier_name = (data element 1, data element 2, …, data
                        element N)
```

From this general syntax, it is clear that a Python tuple can be created by enclosing its comma-separated data elements within parentheses, as demonstrated in Listing 5.21.

Listing 5.21	Examples demonstrating the creation of Tuples

```
# Creating a tuple for crops rotation order at a farm
crop_rotation = ("wheat", "soybean", "corn", "alfalfa")
print("Print Crop Rotation Pattern:", crop_rotation)

# Creating a tuple of tuples containing monthly rainfall data
monthly_rainfall = (("Jan.", 3.2),("Feb.", 2.8), ("Mar.", 4.1))
print("Monthly Rainfall:", monthly_rainfall)

# Creating a tuple of tuples containing farms' yield data
top_yields = (("corn", 7.8), ("wheat", 6.5), ("rice", 6.1))
print("Top Crop Yields:", top_yields)

# Creating a tuple of tuples containing livestock data
livestock_wts = (("Holstein",680), ("Jersey",450),("Angus",700))
print("Livestock with Weights:", livestock_wts)
```

Output of Listing 5.21:

The self-explanatory output of Listing 5.21 is shown below.

```
Print Crop Rotation Pattern: ('wheat', 'soybean', 'corn', 'alfalfa')
Monthly Rainfall: (('Jan.', 3.2), ('Feb.', 2.8), ('Mar.', 4.1))
Top Crop Yields: (('corn', 7.8), ('wheat', 6.5), ('rice', 6.1))
Livestock with Weights: (('Holstein', 680), ('Jersey', 450), ('Angus', 700))
```

5.3.2 Accessing Tuple Data Elements

The ordered collection of data elements is indexed in a Python Tuple starting from the first data element on the left that has index [0], the second data element has index [1], and so on. On the other hand, negative indexing starts from the left (meaning it starts from the end), where −1 refers to the last item, −2 refers to the second last item, and so on. Hence, like Python List, the index operator provided in square brackets (with both positive and negative integers) is used to access data elements or slices of Tuple data elements. The general syntax of accessing an individual Tuple data element is

```
tuple_name[index value of tuple data element/item]
```

The access of Tuple data elements has been demonstrated in Listing 5.22.

Listing 5.22	Examples demonstrating the access to Tuple data items

```
# Yearly (from 2019-2025) crop yield data in order

yearly_crop_yield = (330, 560, 499, 560, 490, 600, 279)
# print 4th    data item
print("The 3rd data item:", yearly_crop_yield[3])

# print data items from 0 to 7 skipping 1 item each time
print("Data items at even indices:", yearly_crop_yield[0:7:2])

# print data items from 3rd to last in tuple
print("Items 3rd to last in tuple:", yearly_crop_yield[2:])

# print first four data items
print("First four data items:", yearly_crop_yield[:3])

# print 3rd data item from right
print("3rd data from right:", yearly_crop_yield[-3])

# print last 3 data items from right
print("Last 3 data items from right:", yearly_crop_yield[-3:])
```

Output of Listing 5.22:

The self-explanatory output of Listing 5.22 is shown below.

The 3rd data item: 560
Data items at even indices: (330, 499, 490, 279)
Items 3rd to last in tuple: (499, 560, 490, 600, 279)
First four data items: (330, 560, 499)
3rd data item from right: 490
Last 3 data items from right: (560, 490, 600)

5.3.3 Common Operations and Tuple Methods

All common operations and methods used with Python lists can also be applied to Tuples in the same way. These include element traversal using a *for* loop, concatenation with the + operator, repetition with the * operator, membership testing using in and not in, unpacking of tuple elements into variables, and built-in functions like len(), min(), max(), and sum(). Additionally, tuple methods such as count() and index() work similarly to how they are used with lists. Therefore, Listing 5.23 demonstrates the use of all these operations with Python Tuples without requiring any additional explanation.

Listing 5.23	Common operations on Tuple data elements

```
print("==========Traversal of Tuple Data Elements==========")
# Tuple data elements traversal using for Loop
inorganic_fertilizers = ("Urea","Ammonium sulfate","Ammonium
nitrate")

# for loop to iterate through tuple data items
for element in inorganic_fertilizers:
  print("Data Item: ", element)

# for loop to iterate tuple data items with index values
for index in range(len(inorganic_fertilizers)):
  print("Data Item [",index, "] =", inorganic_fertilizers[index])

print("=========Membership Testing of Tuple Data Item==========")
# Membership testing (using in or not operators in a Tuple)
field_sensors = ("temperature", "moisture", "pH", "light")
if "pH" in field_sensors:
   print("Send pH reading from field to server")
else:
    print("Ignore the value")

# Counting and searching Methods for tuple
temperature_readings = (22.9, 22.5, 21.8, 22.5, 20.9)

print("======Counting/Searching Data Item in Tuple======")

# Counting the number of occurrences of a data item
total_occurrences = temperature_readings.count(22.5)
print("Number of occurrence of given value:", total_occurrences)

# Finding the first occurrence/position of a data item in tuple
position = temperature_readings.index(22.5)
print("The first location of the given value:", position)

print("======Use of built-in functions with Tuple======")
yearly_yield = (330, 560, 499, 560, 490, 600, 279)

# Print the length of tuple
print("Total items in tuple:", len(yearly_yield))
```

```
# Print the minimum value in tuple
print("Minimum value in tuple:", min(yearly_yield))

# Print the maximum value in tuple
print("Maximum value in tuple:", max(yearly_yield))

# Print the sum of all tuple elements
print("Sum of all tuple values:", sum(yearly_yield))
print("======Use of + and * operators with Tuple======")

# Concatenation of two tuples
fruits = ("apple", "banana", "cherry")
vegetables = ("cucumber", "cauliflower", "cabbage")
fruits_vegetables = fruits + vegetables
print("Concatenated Tuple:", fruits_vegetables)

# creating concatenated copies of a tuple
milk_products = ("yogurt", "butter", "cream")
product_quantity = milk_products * 2
print("Concatenated Tuple Copies:", product_quantity)

print("======Unpack Tuple Elements======")
sensor_val = (("moisture", 30), ("temperature", 25))
(m_val, temp_val) = sensor_val
print("Moisture value:", m_val)
print("Temperature value:", temp_val)
```

Output of Listing 5.23:

The self-explanatory output of Listing 5.23 is shown below.

```
==========Traversal of Tuple Data Elements==========
Data Item: Urea
Data Item: Ammonium sulfate
Data Item: Ammonium nitrate
Data Item [0] = Urea
Data Item [1] = Ammonium sulfate
Data Item [2] = Ammonium nitrate
==========Membership Testing of Tuple Data Item==========
Send pH reading from field to server
======Counting/Searching Data Item in Tuple======
Number of occurrence of given value: 2
The first location of the given value: 1
======Use of built-in functions with Tuple======
Total items in tuple: 7
Minimum value in tuple: 279
Maximum value in tuple: 600
Sum of all tuple values: 3318
======Use of + and * operators with Tuple======
Concatenated Tuple: ('apple', 'banana', 'cherry', 'cucumber', 'cauliflower', 'cabbage')
Concatenated Tuple Copies: ('yogurt', 'butter', 'cream', 'yogurt', 'butter', 'cream')
======Unpack Tuple Elements======
Moisture value: ('moisture', 30)
Temperature value: ('temperature', 25)
```

5.4 Set

In Python, the Set data structure is used to store a collection of unordered, non-duplicate, and unchangeable data elements under the same variable. It is important to keep in mind that although data elements in Python Set are unchangeable, after creation, added data elements can be deleted, and new elements can be added to the set.

5.4.1 Set Creation

In Python, a Set can be created or defined in multiple ways, as shown in Listing 5.24.

Listing 5.24	Examples demonstrating the creation of a Python Set

```
# Way to create an empty set and adding data items later
farm_activities = set()
farm_activities.add("sowing")
farm_activities.add("ploughing")
farm_activities.add("irrigation")
print("Farm activities Set:", farm_activities)

# Way to create a set using curly braces {}
moisture_reading_set = {30, 45, 10}
print("Moisture readings Set:", moisture_reading_set)

# Create a set from a list
livestock_set = set(["cow", "camel", "goats"])
print("Set of livestock:", livestock_set)

# Create a set from a list comprehension
yield_record = set([crop_yield for crop_yield in range(40, 45) if
crop_yield <45 ])
print("Set of yield record:", yield_record)

# Create a set from a tuple
temperature_reading_set = set((18, 5, 11))
print("Temperature readings Set:", temperature_reading_set)

# Create a set using range function
crop_ids = set(range(1, 6))
print("Set of crop IDs:", crop_ids)
```

Output of Listing 5.24:

Possible output after 1st time execution:
Farm activities Set: {'irrigation', 'ploughing', 'sowing'}
Moisture readings Set: {10, 45, 30}
Set of livestock: {'goats', 'cow', 'camel'}
Set of yield record: {40, 41, 42, 43, 44}
Temperature readings Set: {18, 11, 5}
Set of crop IDs: {1, 2, 3, 4, 5}

Possible output after 2nd time execution:
Farm activities Set: {'irrigation', 'ploughing', 'sowing'}
Moisture readings Set: {10, 45, 30}
Set of livestock: {'camel', 'goats', 'cow'}
Set of yield record: {40, 41, 42, 43, 44}
Temperature readings Set: {18, 11, 5}
Set of crop IDs: {1, 2, 3, 4, 5}

Possible output after 3rd time execution:
Farm activities Set: {'ploughing', 'sowing', 'irrigation'}
Moisture readings Set: {10, 45, 30}
Set of livestock: {'cow', 'goats', 'camel'}
Set of yield record: {40, 41, 42, 43, 44}
Temperature readings Set: {18, 11, 5}
Set of crop IDs: {1, 2, 3, 4, 5}

Explanation:
The output of Python sets might be different after each execution as it has been highlighted after three different executions. It is important to note that starting from Python 3.7, the order of set elements often appears to be maintained (or remains the same after each execution), as you can see in the output values of crop_ids, temperature_reading_set, and moisture_reading_set defined in Listing 5.24. Nevertheless, it is essential to remember that sets in Python remain unordered collections, meaning the order of elements is not guaranteed and may change in different executions.

5.4.2 Accessing Set Data Elements

The unordered collection of data elements is not indexed in a Python Set, and therefore you cannot access data items in a set by referring to an index. Nevertheless, using a *for* loop, you can iterate through the set items and by using the *in* (or *not in*) operators, you can check whether a specific value is present in the set or not, as shown in Listing 5.25.

Listing 5.25	Examples demonstrating the membership testing of set data elements

```
crops_and_yield = {"Wheat", 1000, "Rice", 500}

# Iterating Python Set using for loop
for val in crops_and_yield:
    print("Set value:", val)
```

```
# Using "in" operator to check the membership of a set
if 1000 in crops_and_yield:
    print("1000 is the set element")
else:
    print("1000 is not the set element")
```

Output of Listing 5.25:

Possible output after 1st time Execution:

Set value: 1000
Set value: Wheat
Set value: 500
Set value: Rice
1000 is the set element

Possible output after 2nd time Execution:

Set value: 1000
Set value: 500
Set value: Rice
Set value: Wheat
1000 is the set element

5.4.3 Adding Elements to Python Set

The *add()* and *update()* methods allow adding elements to a Python set. The *add()* method is used to insert a single element, while the *update()* method enables adding multiple elements from an iterable. This functionality is demonstrated in Listing 5.26.

Listing 5.26	Examples demonstrating the *add()* and *update()* methods on Set data

```
# Add an individual data element in a Set
farm_equipment = {"tractor", "plow", "seeder", "harvester"}
print("Farm equipment:", farm_equipment)
farm_equipment.add("sprayer")
print("Farm equipment after adding an item:", farm_equipment)

# Add multiple data elements from an iterable (List here)
field_activities = ["cultivar", "sprayer", "mower", "loader"]
farm_equipment.update(field_activities)
print("Farm equipment after adding items:", farm_equipment)
```

Output of Listing 5.26:

Possible output after 1st time Execution:

Farm equipment: {'harvester', 'seeder', 'plow', 'tractor'}
Farm equipment after adding an item: {'harvester', 'seeder', 'tractor', 'plow', 'sprayer'}
Farm equipment after adding items: {'harvester', 'seeder', 'loader', 'tractor', 'plow', 'cultivar', 'mower', 'sprayer'}

Possible output after 2nd time Execution:

Farm equipment: {'harvester', 'seeder', 'tractor', 'plow'}
Farm equipment after adding an item: {'seeder', 'sprayer', 'plow', 'harvester', 'tractor'}
Farm equipment after adding items: {'seeder', 'sprayer', 'mower', 'plow', 'cultivar', 'harvester', 'tractor', 'loader'}

5.4.4 Removing Elements from Python Set

The *remove(), discard()*, *pop()*, and *clear()* methods are used to delete elements from a Python set. The key difference between *remove()* and *discard()* methods is that *remove()* raises an error if the specified element is not found, whereas *discard()* does not. The *pop()* method removes and returns a random element from the set, while the *clear()* method deletes all elements, leaving the set empty, as illustrated in Listing 5.27.

Listing 5.27	Examples demonstrating the removal of Python Set data elements

```
farm_equipment = {"tractor", "plow", "seeder", "sprayer",
                                                    "harvester"}
print("Farm equipment:", farm_equipment)

# Remove a data item from Set using remove() method
farm_equipment.remove("sprayer")
print("After deleting sprayer:", farm_equipment)

# Remove a data item from Set using discard() method
farm_equipment.discard("seeder")
print("After deleting seeder:", farm_equipment)

# Remove a data item from Set using pop() method
farm_equipment.pop()
print("After deleting random item:", farm_equipment)

# Remove a data item from Set using pop() method
farm_equipment.clear()
print("After deleting all items:", farm_equipment)
```

Possible output of Listing 5.27:

Farm equipment: {'seeder', 'tractor', 'harvester', 'plow', 'sprayer'}
After deleting sprayer: {'seeder', 'tractor', 'harvester', 'plow'}
After deleting seeder: {'tractor', 'harvester', 'plow'}
After deleting random item: {'harvester', 'plow'}
After deleting all items: set()

5.4.5 Operations on Python Sets

Standard math's Set operations, i.e., union, intersection, difference, and symmetric difference, can be performed using Python sets as demonstrated in Listing 5.28.

Listing 5.28	Examples demonstrating the use of standard Math Set operations

```
# Flowers grown at farm A and farm B
farm_A_flowers = {"Rose", "Sunflower", "Lotus", "Daffodil"}
farm_B_flowers = {"Lily", "Rose", "Daffodil", "Marigold"}

'''
Union of both farms shows flowers grown at either Farm A or Farm B
(or both)
'''
all_flowers = farm_A_flowers.union(farm_B_flowers)
print("Flowers at both Farm A and B:", all_flowers)

# Intersection of both farms shows flowers common in both farms
common_flowers = farm_A_flowers.intersection(farm_B_flowers)
print("Common Flowers at Both Farms:", common_flowers)

# Difference: flowers grown only at Farm A but not at Farm B
unique_farm_A_flowers = farm_A_flowers.difference(farm_B_flowers)
print("Flowers only at Farm A:", unique_farm_A_flowers)

# Difference: flowers grown only at Farm B but not at Farm A
unique_farm_B_flowers = farm_B_flowers.difference(farm_A_flowers)
print("Flowers only at Farm B:", unique_farm_B_flowers)

# Symmetric Difference:flowers unique to each farm (not in both)
unique_flowers = farm_A_flowers.symmetric_difference(farm_B_flowers)
print("Unique Flowers at Each Farm:", unique_flowers)
```

Possible output of Listing 5.28:

Flowers at both Farm A and B: {'Marigold', 'Sunflower', 'Lily', 'Lotus', 'Rose', 'Daffodil'}
Common Flowers at Both Farms: {'Daffodil', 'Rose'}
Flowers only at Farm A: {'Lotus', 'Sunflower'}
Flowers only at Farm B: {'Marigold', 'Lily'}
Unique Flowers at Each Farm: {'Marigold', 'Sunflower', 'Lily', 'Lotus'}

Explanation:
The union method returns the set containing data elements from both sets.
The intersection method returns the set containing common data elements of both sets.
The difference method returns the set containing only data elements of the first set that are not present in the second set.
The symmetric_difference() method returns the set containing data elements that keep only those elements that are not common to both sets.

5.5 Python Dictionary

In Python, the Dictionary data structure is used to store non-duplicate, changeable (mutable), and ordered key:value pairs. Dictionaries do not allow duplicate keys, meaning each key in a dictionary must be unique. They are changeable and allow modifications such as adding, updating, or removing items after creation. Moreover, dictionaries are ordered, meaning that items maintain a defined sequence and their order remains unchanged. (Note: Since Python 3.7, dictionaries preserve insertion order, whereas in earlier versions, these were unordered.) Python Dictionaries are ideal for scenarios where data elements need to be organized and accessed by identifiers directly or you need to group and analyze data by categories or labels (e.g., grouping yields by crop type).

5.5.1 Python Dictionary Creation and Accessing Data Elements

Python Dictionaries are defined with keys and values enclosed in braces (curly brackets { }). The general syntax of Python dictionary creation is shown as follows:

```
dictionary_name = {
             "Key1": Value1,
             "Key2": Value2,
             ...
             "KeyN": ValueN
}
```

Considering a scenario where an agriculturist is interested in storing the crop yield (values) associated with crop names (keys), the creation of a Python Dictionary has been illustrated in Listing 5.29.

Listing 5.29	Example of Python dictionary creation

```
# Creation of Python Dictionary to store crop yields
crop_yields = {
    "Wheat": 4.5,
    "Rice": 2.2,
    "Maize": 4.7,
    "year": 2024
}
```

5.5.2 Common Dictionary Operations

Considering the *crop_yields* dictionary created in Listing 5.29, accessing and traversing of dictionary keys and values have been explained in Listing 5.30.

Listing 5.30	Examples of accessing and traversing Python Dictionary keys and values

```
# Two ways to get Dictionary Keys
# 1. print all dictionary keys using simple for loop
for var in crop_yields:
  print("Key:", var)

# 2. print all dictionary keys using for loop and keys()
for var in crop_yields.keys():
  print("Key:", var)

# Two ways to get dictionary values
# 1. print dictionary values using square brackets[]
for val in crop_yields:
  print("Crop Yield:", crop_yields[val])

# 2. print dictionary values using values() method
for val in crop_yields.values():
  print("Crop Yield:", val)

# Use of items() method to print both keys and values
for k, v in crop_yields.items():
  print("Key:", k, "& Value:", v)

# Two ways to get value associated with a key
# 1. using key in square brackets []
wheat_yield = crop_yields["Wheat"]
print("Wheat Yield:", wheat_yield)

# 2. using get() method of dictionary
rice_yield = crop_yields.get("Rice")
print("Rice Yield:", rice_yield)
# Printing the yield of a specific crop in a specific year
print("Rice Yield",crop_yields["year"],"=",crop_yields["Rice"])
```

Output of Listing 5.30:

The self-explanatory output of Listing 5.30 is shown below.

Key: Wheat
Key: Rice
Key: Maize
Key: year
Key: Wheat
Key: Rice
Key: Maize
Key: year
Crop Yield: 4.5
Crop Yield: 2.2
Crop Yield: 4.7
Crop Yield: 2024
Crop Yield: 4.5
Crop Yield: 2.2
Crop Yield: 4.7
Crop Yield: 2024
Key: Wheat & Value: 4.5
Key: Rice & Value: 2.2
Key: Maize & Value: 4.7
Key: year & Value: 2024
Wheat Yield: 4.5
Rice Yield: 2.2
Rice Yield 2024 = 2.2

5.5.3 Membership Testing, Adding, and Removal of Dictionary Elements

The membership testing, adding, and removal of dictionary items have been demonstrated in Listing 5.31.

Listing 5.31	Examples of membership testing, adding, and removal of items in Python Dictionary

```
print(crop_yields.items())
# Use of in operator to check the presence of a key
if "Maize" in crop_yields:
  print("Yes, 'Maize' is a key in crop_yields")

# Two ways to add to a dictionary value
# 1. By assigning a value to a dictionary key
crop_yields["climate"] = "tropical"
print("After adding information about climate")
print(crop_yields.items())
```

```
# 2. Using update() method to add a new key-value pair
crop_yields.update({"soil": "sandy"})
print("After adding information about soil type")
print(crop_yields.items())

# Two ways to remove dictionary data items
# 1. Use pop() method to delete a data item
crop_yields.pop("climate")
print("After deleting data associated with 'climate' key")
print(crop_yields.items())

# 2. Use popitem() method to delete the last data item
crop_yields.popitem()
print("After deleting last inserted key:value pair")
print(crop_yields.items())
```

Output of Listing 5.31:

The self-explanatory output of Listing 5.31 is shown below.

```
dict_items([('Wheat', 4.5), ('Rice', 2.2), ('Maize', 4.7), ('year', 2024)])
Yes, 'Maize' is a key in crop_yields
After adding information about climate
dict_items([('Wheat', 4.5), ('Rice', 2.2), ('Maize', 4.7), ('year', 2024), ('climate', 'tropical')])
After adding information about soil type
dict_items([('Wheat', 4.5), ('Rice', 2.2), ('Maize', 4.7), ('year', 2024), ('climate', 'tropical'), ('soil',
'sandy')])
After deleting data associated with 'climate' key
dict_items([('Wheat', 4.5), ('Rice', 2.2), ('Maize', 4.7), ('year', 2024), ('soil', 'sandy')])
After deleting last inserted key:value pair
dict_items([('Wheat', 4.5), ('Rice', 2.2), ('Maize', 4.7), ('year', 2024)])
```

5.6 Exercises

Problem 5.1 An agricultural research institute needs a system to track fixed data on crop varieties, such as harvest season, average yield, and ideal planting conditions. The system should be able to facilitate researchers to

- store information for new crop varieties
- view all crop varieties records
- search specific crop variety and related attributes in stored data

Problem 5.2 A livestock manager requires a computer application to efficiently track the health status of various cow breeds in a cattle ranch. This program should enable the manager to

- categorize cow breeds based on breed type, vaccination status, and sale status, ensuring no duplicate entries.

- add or remove cows from specific categories as needed.
- check common and unique cows across different categories.
- remove an entire category when it is no longer needed, such as clearing the vaccination category once all cows in that group have been vaccinated.

Problem 5.3 An entomologist researcher needs a system to store and manage pest-related data (including species name, active seasons, average crop damage (in percentage), and preferred habitats). The system should be able to

- store information for new pest species in a tuple (since pest attributes are stable, Python Tuples are ideal for storing this data)
- view all pest records
- search for specific pest or pest attributes in stored data

Problem 5.4 A horticulturist requires a system to keep track of essential information (including optimal planting month, harvest month, required soil pH, and yield per hectare) to manage various crops. The system should be able to facilitate the horticulturist to store, access, delete, and update crop information by crop name.. Show the use of Python Dictionary to implement this system.

Problem 5.5 A forestry manager requires a system to monitor various tree species to support sustainable logging, reforestation, and conservation efforts. Each tree species needs to be stored and managed with specific data such as optimal growth conditions, carbon storage capacity, economic value, and common uses. Write a Python program that enables the forestry manager to efficiently perform the following tasks on the data provided in Table 5.3 (using the tree species name as the reference):

- add new data for a tree species
- retrieve specific information about a specific species
- update existing data related to a species
- delete data for a tree species
- iterate through all tree species

Table 5.3 Tree species data

Tree name	Growth rate	Wood density	Carbon storage capacity (kg/tree)	Timber value (per m3)	Common usage
Oak	Medium	0.75	500	150	Furniture, flooring, construction
Pine	Fast	0.4	300	100	Paper, furniture, construction
Maple	Slow	0.6	400	180	Cabinetry, flooring, decorative items

Chapter 6
File Handling in Python

6.1 File Handling

File handling is one of the core aspects of any computer programming language including Python. The use of file handling allows programmers to develop applications related to data retrieval, manipulation, and storage. Considering the nature of data, computer files can be broadly classified into two main categories, i.e., text files and binary files. The characteristics of both types of files have been mentioned in Table 6.1.

Note: It is important to note that the distinction between text and binary files is conceptual, as computing devices store all files as a sequence of bits (0s and 1s). The way how stored bits are interpreted by computer programs makes them different. Text files are treated differently by the use of standard encoding schemes that convert stored bytes into readable characters.

6.2 File Operations

Typical operations that can be performed with (text and binary) files are

- File Existence Check
- File Opening and Closing
- File Reading
- File Writing
- File Deleting
- File Closing

M. A. Iqbal, *Python for Agriculturists*,
https://doi.org/10.1007/978-3-032-01442-9_6

Table 6.1 Differences between text files and binary files

Text files	Binary files
Contain data in human-readable formats	Contain data in machine-readable formats
Simple text editors can be used to open, read, and write data	Text editors are unable to interpret data so specialized software is required to open, check, and manipulate data in these files
Each character in text files is stored as a specific sequence of bytes (typically using the ASCII or UTF-8 encoding scheme)	Binary files store data in the raw form and based on specific file formats, the correct meanings of bytes are interpreted
Examples are the files stored with extensions such as .doc(x), .txt, .csv, .json, etc.	Examples are the files stored with extensions such as .jpg, .dat, .mp3, .zip, .exe, etc.

6.2.1 File Existence Check

Before reading (or writing) text or binary files, it is better to check the existence of that file on that system. File existence can be checked using the *os.path* or *pathlib* modules. There are two methods, i.e., *exists()* and *isfile()*, in *os.path* module that are used for this purpose. The key difference between these two is that *exists()* method returns *True* if the given path to a folder or a file exists, whereas *isfile()* returns *True* only if the given path is a path to a file and not a folder. The use of these methods is shown in Listing 6.1.

Listing 6.1	Code to check the existence of a file (or folder)

```
# Checking file existence using isfile() method
import os.path
if os.path.isfile("FarmData.txt"):
    print("FarmData.txt exists")

# Checking file or folder existence using exists() method
import os.path
if os.path.exists("FarmData.txt"):
    print("FarmData.txt exists")
```

Here, both built-in methods, i.e., *isfile("FarmData.txt") and exists("FarmData.txt")* return *True* if there exists a file named *FarmData.txt* in the current directory. If the file is available in another directory or folder, then you can use the absolute address with the file name. For example, suppose this file named *FarmData.txt* is stored in a directory or folder named *DigAgri* on the *D Drive* of the Windows operating system, then the absolute path can be used in the open method using two ways as shown in Listings 6.2 and 6.3.

Listing 6.2	Code to check the existence of a file (or folder) using an absolute path *with prefix "r"*

```
import os.path
if os.path.isfile(r"D:\DigAgri\FarmData.txt"):
    print("FarmData.txt exists")
```

Here, the prefix "*r*" before an absolute filename string indicates that the string is a raw string. The use of this prefix ensures that backslashes are interpreted as literal characters. Without this prefix, you can use escape sequences to represent backslashes properly, as shown in Listing 6.3.

Listing 6.3 Code to check the existence of a file (or folder) using an absolute path *without prefix "r"*

```
import os.path
if os.path.isfile("D:\\DigAgri\\FarmData.txt"):
    print("FarmData.txt exists")
```

6.2.2 File Opening and Closing in Python

Opening a file is required to make the file available to perform other file handling operations, i.e., reading file content, writing data in the file, and editing or modifying existing data or file content. Ensuring that a file is properly closed (after performing required operations) is equally important, as it prevents data corruption and system resource leaks (which refers to a state where a computer program fails to release system resources (i.e., file, network connection, memory, etc.) it has acquired). Thus, file opening and closing are the fundamental operations to be performed before any other type of interaction with a file, i.e., reading or writing file data.

In Python, the built-in *open()* function can be used in two ways: using a *with* statement or without it to open a file. The basic syntax of using these two ways is shown as follows:

```
(1) variable_name = open("file_name.extension", "mode")
(2) with open("file_name.extension", "mode") as variable_name:
```

The key difference between these two file-opening ways is that when using the *with* statement, the file closes automatically. This is the main reason for recommending this way for file opening. On the other hand, when opening a file with the *open()* function without the *with* statement, the file must be explicitly closed using the *close()* function. However, it becomes evident that in both cases, you must declare a variable (*variable_name* as shown in the syntax above) on which you can invoke file handling functions to perform file handling operations. The "*file_name.extension*" parameter is the string that represents the file to be opened (including its path if the file is not stored in the same directory), and *mode* is the file access mode to perform further operations, i.e., read, write, and append. Considering the specific nature of various file handling operations, Python provides different file modes as shown in Table 6.2.

As mentioned earlier, when a file is opened using the *open()* function in Python, it should be explicitly closed after completing the necessary operations. This helps prevent data corruption and avoids unnecessary use of system resources. To close an open file, the built-in *close()* function is used. A recommended way is to use an exception handling mechanism to open and close files in Python, as it ensures that files are closed properly even in case of runtime errors, as shown in Listing 6.4.

Table 6.2 Type and description of Python file handling modes

Mode	Description
"r"	(Default) Read-only mode for text files. File must exist to open in read-only mode otherwise an error will be raised
"rb"	Read-only mode for binary files. File must exist to open in read-only mode otherwise an error will be raised
"r+"	Opening file for both reading and writing. File must exist otherwise an error will be raised
"w"	Write mode for text file. If file does not exist then a new file will be created. Otherwise, contents of the already existing file will be overwritten
"wb"	Write mode for binary files. If file does not exist then a new file will be created. Otherwise, contents of the already existing file will be overwritten
w+	Opening a text file for both writing and reading. A new file will be created (that will override the file if the file with same name already exists)
wb+	Opening a binary file for both writing and reading. A new file will be created (that will override the file if the file with same name already exists)
"a"	Opening a text file in append mode. Data will be added at the end of the existing file. Otherwise, new data will be written to newly created file
"ab"	Opening a text file in append mode. Data will be added at the end of the existing file. Otherwise, new data will be written to newly created file
"a+"	Opening a text file for both reading and appending. Data will be added at the end of the existing file. Otherwise, new data will be written to newly created file
ab+	Opening a binary file for both reading and appending. A new file will be created (that will override the file) if the file with the same name already exists
x	Exclusive text file creation for reading mode. Creates a new file and raises an error if the file already exists
xb	Exclusive binary file creation for reading mode. Creates a new file and raises an error if the file already exists
x+	Exclusive text file creation with read and write modes. Creates a new file and raises an error if the file already exists
xb+	Exclusive binary file creation with read and write in binary mode. Creates a new binary file for reading and writing and raises an error if the file already exists

Listing 6.4	File operations with exception handling

```
source_file = None
try:
    source_file = open("FarmData.txt", "r")
except FileNotFoundError:
    print("Error: Source file not found")
except PermissionError:
    print("Error: Permission denied to read this file")
except Exception:
    print("An unpredicted error occurs")
else:
    pass
    # Write file processing programming logic here

finally:
    if source_file != None:
        source_file.close()
```

This code in Listing 6.4 uses a *try-except-else-finally* block to handle possible errors (regarding the opening and closing of a file) and ultimately ensures proper resource management. If the file is not found, a *FileNotFoundError* is raised and caught by an *except* handler block to display an appropriate error message. If the program lacks permission to access the file, then a *PermissionError* is raised and caught by an *except* handler block to display a different relevant message. Any other unexpected errors are caught by a general *Exception* handler. If no exception occurs, the *else* block is reached, where the actual file processing logic is recommended to be placed. Regardless of whether an error occurs or not, the *finally* block ensures that the file is closed. Although the basics of the exception handling mechanism using only the *try-except* block have been briefly discussed in Sect. 2.2.2 of Chap. 2, additional details about the use of the full *try-except-else-finally* structure are mentioned as follows:

- The *try* block is used to test a section of code for potential errors.
- If an error occurs, the single or multiple *except* block(s) allow(s) the programmer to handle the raised exception appropriately (in the respective *except* block).
- In case no error is raised in the *try* block, the code inside the *else* block will be executed.
- The *finally* block contains code that will run regardless of an exception is raised or not and whether it is handled or not.

6.2.3 Text File Reading in Python

Reading a text file allows the programmer to access the file content for further processing. After file opening in any of the read modes (i.e., "r, or "r+"), you can use various built-in methods to read the content of a file, i.e., the *read()* method to read the entire content of a file and *read(line)* method to read file content line by line. Listing 6.5 demonstrates how to use the *read()* method and Listings 6.6, 6.7, and 6.8 illustrate line-by-line reading ways of a text file named *CropYieldData.txt*, which contains the following sample data.

CropYieldData.txt

Year	Crop	Yield
2020	Wheat	9.1
2021	Rice	7.5
2023	Oat	3.5
2024	Barley	5.9

Listing 6.5	Example: Reading the entire file using the *read()* method

```
source_file = None
try:
    # Open the file in read mode
    source_file = open("CropYieldData.txt", "r")
except FileNotFoundError:
    print("Error: Source file not found")
except PermissionError:
    print("Error: Permission denied to read this file")
except Exception:
    print("An unpredicted error occurs")
else:
    # Read the entire content of the file
    content = source_file.read()
    # Print the content
    print(content)

finally:
    if source_file != None:
        source_file.close()
```

Output of Listing 6.5:

The self-explanatory output of Listing 6.5 is shown below.

```
Year  Crop   Yield
2020  Wheat  9.1
2021  Rice   7.5
2023  Oat    3.5
2024  Barley 5.9
```

For text files containing a large amount of data, it is impractical to read the entire content of the file, and for this reason, it is recommended to use the *readline()* method with a *while* loop to read the file data line by line.

Listing 6.6	Example: Reading the entire file line-by-line using the *readline()* method

```
source_file = None
try:
    # Open the file in read mode
    source_file = open("CropYieldData.txt", "r")
except FileNotFoundError:
    print("Error: Source file not found")
except PermissionError:
    print("Error: Permission denied to read this file")
except Exception:
    print("An unpredicted error occurs")
else:
    # Read the first line
    a_line = source_file.readline()
    while a_line:
        print(a_line)
        # Read the next line
        a_line = source_file.readline()

finally:
    if source_file != None:
        source_file.close()
```

Output of Listing 6.6:

```
Year   Crop    Yield
2020   Wheat   9.1
2021   Rice    7.5
2023   Oat     3.5
2024   Barley  5.9
```

The output of this file execution shows that there is a gap between lines that is due to the reading of the new line character in each line. To remove this new line character, you can use the *strip()* function of the string class. To get output correctly, the code line *print(a_line)* should be replaced with print(*a_line.strip()*), as shown in Listing 6.7.

Listing 6.7	Example: Reading the entire file line-by-line using the *strip()* method

```
source_file = None
try:
    # Open the file in read mode
    source_file = open("CropYieldData.txt", "r")
except FileNotFoundError:
    print("Error: Source file not found")
except PermissionError:
    print("Error: Permission denied to read this file")
except Exception:
    print("An unpredicted error occurs")
else:
    # Read the first line
    a_line = source_file.readline()
    while a_line:
        print(a_line.strip())
        # Read the next line
        a_line = source_file.readline()
finally:
    if source_file != None:
        source_file.close()
```

Output of Listing 6.7:

```
Year  Crop    Yield
2020  Wheat   9.1
2021  Rice    7.5
2023  Oat     3.5
2024  Barley  5.9
```

Another way of reading a file line-by-line without using the *readline()* method is through the use of a *for* loop. For example, the above implementation can be done by using Python's *for* loop to read and print the file content line by line.

Listing 6.8	Example: Reading the entire file line-by-line using the simple *for* loop

```
source_file = None
try:
    # Open the file in read mode
    source_file = open("CropYieldData.txt", "r")
except FileNotFoundError:
    print("Error: Source file not found")
except PermissionError:
    print("Error: Permission denied to read this file")
except Exception:
    print("An unpredicted error occurs")
else:
```

```
    # Read each line one by one
    for line in source_file:
        print(line.strip())

finally:
    if source_file != None:
        source_file.close()
```

6.2.4 Reading Binary Files in Python

As already mentioned, binary files' data format is not intended to be accessed and processed as text. Therefore, there is no *readline()* method to read a binary file. However, the *read()* method can be used in two different ways to read data from a binary file. Listing 6.9 demonstrates how to use the *read()* method (without parameters) to read the entire binary file (named *AgriFieldImage.png*) at once. Listing 6.10 illustrates the use of the *read(integer_chunk_size)* method to read a specific chunk of a binary file (named *AgriFieldImage.png*). Listing 6.11 illustrates the use of the *read(integer_chunk_size)* method to read the entire binary file (named *AgriFieldImage.png*) in chunks. The binary file named *AgriFieldImage.png* is shown in Fig. 6.1.

Listing 6.9	Example: *read()* method to read entire content of a binary file

```
source_file = None
try:
    # Use read binary (rb) mode to open binary file
    source_file = open("AgriFieldImage.png", "rb")
except FileNotFoundError:
    print("Error: Source file not found")
except PermissionError:
    print("Error: Permission denied to read this file")
except Exception:
    print("An unpredicted error occurs")
else:
    # Read the entire content of the file
    file_content = source_file.read()
    # Display the file content (that will be in hex bytes)
    print(file_content)
finally:
    if source_file != None:
        source_file.close()
```

Output of Listing 6.9:

The output of Listing 6.9 will be the file content in hexadecimal bytes.

Fig. 6.1 AgriFieldImage.png

Listing 6.10	Example: *read(integer_chunk_size)* method to read a specific chunk of a binary file

```
source_file = None
try:
    # Use read binary (rb) mode to open binary file
    source_file = open("AgriFieldImage.png", "rb")
except FileNotFoundError:
    print("Error: Source file not found")
except PermissionError:
    print("Error: Permission denied to read this file")
except Exception:
    print("An unpredicted error occurs")
else:
    # Read the entire content of the file
    file_content = source_file.read(5)
    # Display the file content (that will be in hex bytes)
    print(file_content)
finally:
    if source_file != None:
        source_file.close()
```

Output of Listing 6.10:

The output of Listing 6.10 will be the first 5 bytes of the file in hexadecimal format.

Listing 6.11 Example: *read(integer_chunk_size)* method to read the entire binary file in chunks

```
source_file = None
try:
    # Use read binary (rb) mode to open binary file
    source_file = open("AgriFieldImage.png", "rb")
except FileNotFoundError:
    print("Error: Source file not found")
except PermissionError:
    print("Error: Permission denied to read this file")
except Exception:
    print("An unpredicted error occurs")
else:
    while True:
        chunk = source_file.read(1024)
        if not chunk:
            break # End of file
        # Process the chunk (e.g., print or write)
        print(chunk)
finally:
    if source_file != None:
        source_file.close()
```

Output of Listing 6.11:

The output of Listing 6.11 will be the chunk-by-chunk information of the entire binary file.

6.2.5 File Writing in Python

To write text file content, Python provides two built-in methods, i.e., *write()* to write a specified string to a file and *writelines()* to write a specific list of strings to a file. However, before writing, it is recommended to use the *writeable()* method to check whether the contents can be written in a file or not, as shown in Listing 6.12.

Listing 6.12	Example: *Use of writeable(), write() and writelines() methods to write text data in file*

```
source_file = None
try:
    # Use write mode to open a text file
    source_file = open("AgriBranches.txt", "w")
except FileNotFoundError:
    print("Error: Source file not found")
except PermissionError:
    print("Error: Permission denied to read this file")
except Exception:
    print("An unpredicted error occurs")
else:
    if source_file.writable():
        source_file.write("Branches of Agriculture: \n")
        source_file.writelines(["Agronomy \n", "Entomology \n"]
    print("File contents have been written successfully")
finally:
    if source_file != None:
        source_file.close()
```

Output of Listing 6.12:

The output of Listing 6.12 will be the creation of file named AgriBranches.txt. The content of AgriBranches.txt is shown below.
Branches of Agriculture:
Agronomy
Entomology

6.2.6 File Deletion in Python

Depending on the nature of the deleting requirements, Python provides four built-in methods (i.e., *os.remove()*, *send2trash()*, *os.rmdir()*, and *rmtree()*), which can be used to delete files or folders. However, before deleting a file or a folder, it is recommended to check the existence of that file and folder, as shown in Listing 6.13.

Listing 6.13	Code snippet to check the existence of a file

```
import os

file_name = "CropYieldData1.txt"
if os.path.exists(file_name):
    print(f'The file "{file_name}" exists.')
else:
    print(f'The file "{file_name}" does not exist.')
```

6.2.6.1 Deleting File Using *os.remove()* Method

To delete a file, after importing the *os* library, use the *os.remove()* method as shown in Listing 6.14.

Listing 6.14	Code snippet showing the permanent deletion of a file using the *os.remove()* method

```
import os

file_name = "CropYieldData1.txt"
if os.path.exists(file_name):
    print(f'The file "{file_name}" exists.')
    os.remove(file_name)
    print(f'The file "{file_name}" has been successfully deleted.')
else:
    print(f'The file "{file_name}" does not exist.')
```

6.2.6.2 Deleting File Using *send2trash()* Method

Unlike the *os.remove()* method, which permanently deletes the file from the system, the *send2trash()* method of the *send2trash* module will move the file to the recycle bin. The use of the *send2trash()* method has been shown in the Listing.

Listing 6.15	Code snippet showing the deletion of a file to the recycle bin using the *send2trash* method

```
from send2trash import send2trash
file_name = 'CropYieldData1.txt'

try:
    send2trash(file_name)
    print(f"File '{file_name}' sent to trash successfully.")
except FileNotFoundError:
    print(f"File '{file_name}' not found.")
```

6.2.6.3 Deleting Empty Directory Using *os.rmdir()* Method

The *os.rmdir()* method is used to remove an empty directory or folder specified by the given path. The *os.rmdir()* method will raise an exception if the directory or folder is not empty. The use of *os.rmdir()* method has been shown in Listing 6.16.

Listing 6.16	Code snippet showing the deletion of an empty folder

```
import os
directory_name = "SmartPrecFarming"
try:
    os.rmdir(directory_name)
    print(f'Directory "{directory_name}" is successfully
deleted.'
except OSError as e:
    print(f'Error: {e}')
```

6.2.6.4 Deleting Non-empty Directory Using *rmtree()* Method

The *rmtree()* method of the *shutil* module is recommended to use when there is a need to delete an entire directory and its contents, as demonstrated in Listing 6.17.

Listing 6.17	Code snippet showing the deletion of a non-empty folder with its contents

```
import shutil

directory_name = 'SmartPrecFarming'
try:
    shutil.rmtree(directory_name)
    print(f"Directory '{directory_name}' with contents deleted.")
except FileNotFoundError:
    print(f"Directory '{directory_name}' not found.")
```

6.3 Examples: File Handling in Agricultural Context

6.3.1 Example 1: Reviewing Crop Yield Data

Write a Python program for an agronomist who is interested in seeing the record of crop yield data that is stored in a small file named *CropYieldData.txt*.

Solution:
As it is mentioned that the file is of small size, the *read()* method can be used to read the entire file at once, as shown in Listing 6.18.

Listing 6.18	Solution of Example 1

```
with open("CropYieldData.txt", "r") as file:
    content = file.read() # Read the entire file
    print(content)
```

6.3.2 Example 2: Storing Rainfall Data

Write a Python program for an agriculturist who is interested in storing rainfall data for the last 5 weekdays in a text file named *RainfallData.txt*.

Solution:
As the data is already available, after opening the file in write (w) mode, *write()* method can be used to store data in a text file as shown in Listing 6.19.

Listing 6.19	Solution of Example 2

```
with open("RainfallData.txt", "w") as my_file:
    my_file.write("---------------------\n")
    my_file.write("Rainfall Data Report \n")
    my_file.write("---------------------\n")
    my_file.write("Date              Rainfall(mm)        \n")
    my_file.write("2025-01-01            15              \n")
    my_file.write("2025-01-02            12              \n")
    my_file.write("2025-01-03            13              \n")
    my_file.write("2025-01-04            14              \n")
    my_file.write("2025-01-05            15              \n")
```

Output of Listing 6.19:

The output of Listing 6.19 will be the creation of a file (if that file does not already exist) named *RainfallData.txt*. After creation, the following content will be written as shown below.

```
---------------------
Rainfall Data Report
---------------------
Date Rainfall (mm)
2025-01-01    15
2025-01-02    12
2025-01-03    13
2025-01-04    14
2025-01-05    15
```

6.3.3 Example 3: Appending New Records in Rainfall Data

Write a Python program that allows the agriculturist to enter newly obtained data into the text file named *RainfallData.txt*. Moreover, after appending the data, he is also interested in seeing all the stored data in that file.

Solution:

After opening the file in append and read (*a*+) mode, *write()* method can be used to store data in a text file. To read and display the content of this file properly after appending data, it is important to use the *seek()* method before using the *read()* method. The *seek()* method is used to set the file handler position in a file stream. Therefore, in this implementation, *seek()* method has been used to move the position of the file handler to the start of the file, as shown in Listing 6.20.

Listing 6.20	Solution of Example 3

```
with open("RainfallData.txt", "a+") as my_file:
    my_file.write("2025-01-06            13           \n")
    my_file.write("2025-01-07             5           \n")
    my_file.seek(0)
    data = my_file.read()
    print(data)
```

Output of Listing 6.20:

After executing the code in Listing 6.20, the updated content of the file *RainfallData.txt* will be as follows. You can see the last two rows appended at the end.

```
--------------------
Rainfall Data Report
--------------------
Date Rainfall (mm)
2025-01-01    15
2025-01-02    15
2025-01-03    15
2025-01-04    15
2025-01-05    15
2025-01-06    13
2025-01-07     5
```

6.3.4 Example 4: Copying of Soil Test Records

Copying of Text Files: Write a Python program for a soil scientist who wants to copy the data of a text file (named *SoilTest.txt*) containing soil test results to another file (named *FinalSoilTestResults.txt*) and show the progress of copying data in terms of characters and lines copied.

Listing 6.21	Solution of Example 4

```
source_file = open("SoilTest.txt", "r")
destination_file = open("FinalSoilTestResults.txt", "w")

total_lines = total_chars = 0

for line in source_file:
    total_lines += 1
    total_chars += len(line)
    destination_file.write(line)
    print(total_lines, "lines and", total_chars, "chars copied")

source_file.close() # Close the source file
destination_file.close() # Close the destination file
```

Output of Listing 6.21:

The self-explanatory output of Listing 6.21 is shown below.

1 lines and 41 chars copied
2 lines and 61 chars copied
3 lines and 102 chars copied
4 lines and 132 chars copied
5 lines and 147 chars copied
6 lines and 164 chars copied
7 lines and 179 chars copied
8 lines and 197 chars copied

There are multiple ways to improve Python file handling programs. For example, Listings 6.22 and 6.23 are the improved versions of the implementation shown in Listing 6.21.

Listing 6.22	Improved version of the Listing 6.21 implementation showing the progress of copying data in percentage

```
source_file = open("SoilTest.txt", "r")
destination_file = open("FinalSoilTestResults.txt", "w")

current_chars = total_chars = 0
for line in source_file:
    total_chars += len(line)

source_file.seek(0)
for line in source_file:
    current_chars += len(line)
    destination_file.write(line)
    percentage_copied = current_chars/total_chars * 100
    print(round(percentage_copied), " % of file has been copied")
source_file.close()
destination_file.close()
```

Output of Listing 6.22:

The self-explanatory output of Listing 6.22 is shown below.

21% of file has been copied
31% of file has been copied
52% of file has been copied
67% of file has been copied
75% of file has been copied
83% of file has been copied
91% of file has been copied
100% of file has been copied

Listing 6.23	Improved version of the Listing 6.22 with exception handling

```
source_file = None
destination_file = None
try:
    source_file = open("SoilTest.txt", "r")
except FileNotFoundError:
    print("Error: Source file not found")

except PermissionError:
    print("Error: Permission denied to read this file")
else:
    destination_file = open("FinalSoilTestResults.txt", "w")
    current_chars = total_chars = 0

    for line in source_file:
        total_chars += len(line)

    source_file.seek(0)
    for line in source_file:
        current_chars += len(line)
        destination_file.write(line)
        percentage_copied = current_chars/total_chars * 100
        print(round(percentage_copied), "% file is copied")

finally:
    if source_file != None and destination_file != None:
        source_file.close()
        destination_file.close()
```

6.3.5 Example 5: Copying of Binary Files

Copying of Binary Files: Write Python programs to show how images can be copied from one location to another. In Python, there are multiple ways (shown from Listing 6.24 to 6.29) of writing this type of program for copying binary files.

6.3.5.1 Example 1: Copying of Small Binary Files

In Python, the simple and easy way of copying small binary files is possible through the use of *open()*, *read()*, and *write()* functions as shown in Listing 6.24.

Listing 6.24	Python program for Copying Small Binary Files

```
with open("AgriFieldImage.png", "rb") as source,
                  open("CopiedAgriFieldImage4.png", "wb") as dest:
     dest.write(source.read())
```

Large binary files can be copied using the open(), read(), and write() functions by processing the data in chunks as illustrated in Listings 6.25 and 6.26.

6.3.5.2 Example 2: Copying of Large Binary Files

Listing 6.25	Python program for copying large binary files

```
source_file = "AgriFieldImage.png"
dest_file = "CopiedAgriFieldImage.png"

with open(source_file, "rb") as src, open(dest_file, "wb") as dest:
'''
Use of walrus operator= allows assigning a value to a variable
within an expression
'''
   # combining assignment and use in a single step
    while chunk = src.read(1024): # Read 1024 bytes at a time
        dest.write(chunk)

print("File copied without built-in functions.")
```

Listing 6.26	Improved version of implementing Listing 6.25

```
import os

source_file_name = "AgriFieldImage.png"

source_file = open(source_file_name, "rb")
destination_file = open("CopiedAgriFieldImage.png", "wb")

total_size = os.path.getsize(source_file_name)

data_chunk_size = 1024
copied_size = 0

while True:
    chunk = source_file.read(data_chunk_size)
    if not chunk:
        break # Stop when there is no more data to read

    destination_file.write(chunk)

    # Update copied size
    copied_size += len(chunk)

    # Calculate and display progress percentage
    progress = (copied_size / total_size) * 100
    print(f"Copied: {progress:.2f}%")

print("File copied successfully!")

source_file.close()
destination_file.close()
```

6.3.5.3 Example 3: Use of *copyfile()*, *copy()*, and *copy2()* methods

Python also provides high-level methods for copying binary files; for example, using *copyfile()*, *copy()*, and *copy2()* methods of the *shutil* module as shown in Listings 6.27–6.29.

Listing 6.27	File copying using *copyfile()* method

```
import shutil

shutil.copyfile("AgriFieldImage.png", "CopiedAgriFieldImage2.png")
print("File copied using built-in function copyfile().")
```

Listing 6.28	File copying using *copy()* method

```
import shutil

shutil.copy("AgriFieldImage.png", "CopiedAgriFieldImage2.png")
print("File copied using built-in function copy().")
```

Listing 6.29	File copying using *copy2()* method

```
import shutil

shutil.copy2("AgriFieldImage.png", "CopiedAgriFieldImage3.png")
print("File copied using built-in function copy2().")
```

The differences of *copyfile()*, *copy()*, and *copy2()* methods are

- *copyfile()* method simply copies the content of a file to another file
- *copy()* method copies file content along with permissions metadata
- *copy2()* method copies file content and preserves all metadata associated with the file.

6.4 Exercises

Problem 6.1 Write a Python program for a horticulturist who wants to maintain daily growth records of different plant species in a greenhouse. Each record contains the plant name, date, height in cm, and additional remarks. The horticulturist wants to append new records to the text file on a daily basis, along with reading and displaying of stored plant growth data.

Problem 6.2 A large-scale dairy farm maintains health records for hundreds of cattle. Each cow is equipped with a smart collar that records health metrics such as *body temperature, heart rate, and milk production*. These readings are stored in text log files daily. The farm manager is interested in

- Reading animal health records from log files.
- Identifying sick animals based on abnormal readings.

Write a Python program to facilitate the farm manager in this regard.

Problem 6.3 Write a Python program to help a food technologist who needs a system to record quality test results of dairy products in CSV format, including parameters such as pH level, fat percentage, and microbial count. Along with the storage of these records, the technologist wants to automate the reading and displaying of these records.

Problem 6.4 Write a Python program for an entomologist studying the population trends of historical data of different insect species in a region. The data is collected weekly and stored in a text file, including the insect species name, number of sightings, and location.

Chapter 7
Data Science and Python Packages

7.1 Introduction to Data Science

In recent times, Data is one of the most valuable assets across all industries, including agriculture. Its effective use supports better decision-making and helps businesses increase revenue. Extracting meaningful insights from data to make accurate predictions requires the application of Data Science. Data science is a multidisciplinary field mainly focused on analyzing and interpreting data (after basic steps of collection and preprocessing). It involves exploratory data analysis, statistical and mathematical modeling, machine and deep learning methods, and data visualization to extract useful information from available data for actionable insights. From this brief explanation of data science, it is clear that both data analysis and visualization play a central role in data science. To support these tasks, Python offers several built-in and open-source packages and libraries that provide reusable code and make analysis and visualization more efficient and accessible.

7.2 Python Packages

When developing large-scale applications, managing numerous modules can become challenging. To address this issue, Python allows grouping-related modules using packages. A Python package is essentially a directory that contains a special *__init__.py* file along with multiple modules (and sub-packages). The presence of the *__init__.py* file is essential and indicates that the directory is a package and can contain initialization code for the package. Python has several built-in open-source packages that provide reusable code to make development faster and easier. Popular examples of Python packages for data science include:

M. A. Iqbal, *Python for Agriculturists*,
https://doi.org/10.1007/978-3-032-01442-9_7

- *NumPy*: a basic package for numerical operations and scientific computing.
- *Pandas*: a fundamental package for data manipulation, analysis, and handling of tabular data.

7.3 Python *NumPy* Package

NumPy (Numerical Python) is an open-source Python package created by Travis Oliphant in 2005 that provides support to perform mathematical operations on large matrices (multidimensional arrays or lists) efficiently. It is the foundation for most scientific computing tasks in Python and is widely used in fields such as data science, machine learning, and engineering. NumPy offers fast operations such as element-wise calculations. In the context of agriculture, NumPy can be particularly useful for analyzing large datasets (i.e., soil sensor readings, climate measurements, crop yield records, etc.) and enabling agricultural scientists to perform efficient numerical computations.

7.3.1 Benefits of Python NumPy *Package*

The three main benefits of using NumPy are as follows.

7.3.1.1 Memory Efficient

NumPy matrices consume significantly less memory than traditional Python lists because they store elements of the same data type in a compact and efficient format, as demonstrated in Listing 7.1.

Listing 7.1	Example demonstrating memory efficiency of NumPy array over Python List

```
# numpy package is required to be imported to use NumPy arrays
import numpy as np
import sys

# List creation of 1000 elements (from 0-999)
list = range(1000)

# Size of created list
print("Created List Size", sys.getsizeof(int())*len(list))

# Creation of NumPy array of 1000 elements (from 0-999)
nparr = np.arange(1000)

# Size of created NumPy array
print("Created NumPy Array Size", nparr.size * nparr.itemsize)
```

Output of Listing 7.1:

Created List Size 28,000
Created NumPy Array Size 8000

Explanation:
From this example, it becomes clear that NumPy array elements consume less memory compared to the same number of elements stored in a Python List.

7.3.1.2 Fast

NumPy operations are implemented in C, making numerical computations much faster compared to native Python loops or list-based operations, as illustrated in Listing 7.2.

Listing 7.2	Example demonstrating the memory efficiency of NumPy array over Python List

```
# Example to show the time efficiency of using a NumPy array

import numpy as np
import time

SIZE = 1000000
list1 = range(SIZE)
list2 = range(SIZE)

nparr1 = np.arange(SIZE)
nparr2 = np.arange(SIZE)

# Recording time (in seconds) at this point in program
start = time.time()

# Sum of each pair containing items from both lists
result = [(x+y) for x,y in zip(list1, list2)]

# Recording time (in seconds) at this point in program
end = time.time()

# Calculating the total time for above computation
print("List Proc. Time:", round((end - start), 3), "Sec.")

# Recording time (in seconds) at this point in program
start = time.time()

# Sum of each pair containing items from both NumPy arrays
result = nparr1 + nparr2

# Recording time (in seconds) at this point in program
end = time.time()

# Calculating the total time for above computation
print("NumPy Arr. Proc. Time:", round((end - start), 3), "Sec")
```

Output of Listing 7.2:

List Proc. Time: 0.054 Sec.
NumPy Arr. Proc. Time: 0.006 Sec.

7.3.1.3 Convenient

NumPy offers a comprehensive set of built-in functions for mathematical and statistical operations, enabling easy execution of matrix operations and complex calculations on large datasets with minimal coding, as illustrated in examples provided in this chapter.

7.3.2 Benefits of Using NumPy for Agricultural Data

It is important to mention here that the use of NumPy is very helpful in agriculture for handling and analyzing large sets of agricultural data quickly and efficiently. Some common uses include:

- Analyzing crop yields by calculating the average or total yield from different farms, regions, or seasons.
- Analyzing rainfall from different areas to spot signs of drought.
- Comparing soil properties, i.e., pH, nitrogen, phosphorus, and potassium levels across multiple fields.
- Managing sensor data collected from farms, such as temperature, humidity, and soil moisture.
- Using statistical methods to help predict pest attacks based on weather and crop conditions.

The following subsections demonstrate the use of the NumPy package through examples drawn from agricultural scenarios.

7.3.2.1 Creation of NumPy Arrays

Example 7.1 Create a NumPy (one-dimensional) array containing crop yield data over the last 5 years.

Listing 7.3	Creation and displaying of one-dimensional NumPy arrays

```
# Example creating a one-dimensional NumPy array

import numpy as np

crop_yield = np.array([5.3, 4.7, 6.3, 5.8, 3.9])
print("5-Year Crop Yield", crop_yield)

'''
To display year-wise crop yield, the enumerate() function is used
which takes an iterable as input and adds a counter to each
element. It returns an enumerate object, where each item is a
pair containing the counter (index) and the corresponding element
from the iterable. This counter can be used as an index to easily
reference or access each element later when needed.
'''

for index, cyield in enumerate(crop_yield, start = 1):
    print(f"Crop Yield in Year {index}: is {cyield}")
```

Output of Listing 7.3:

The self-explanatory output of Listing 7.3 is shown below.

5-Year Crop Yield [5.3 4.7 6.3 5.8 3.9]

Crop Yield in Year 1: is 5.3
Crop Yield in Year 2: is 4.7
Crop Yield in Year 3: is 6.3
Crop Yield in Year 4: is 5.8
Crop Yield in Year 5: is 3.9

Example 7.2 Create a NumPy (two-dimensional) array containing data of fertilizer usage (N, P, K) across three agricultural fields.

Listing 7.4	Creation and displaying of two-dimensional NumPy arrays

```
'''
Example creating two-dimensional NumPy array Matrix containing
data of Fertilizer use (1 = N, 2 = P, 3 = K in Kgs) across
3 fields
'''
fertilizer_amount = np.array([
    [40, 20, 30],    # Field 1
    [35, 25, 32],    # Field 2
    [38, 22, 34]     # Field 3
])
```

```
'''
To display fertilizer amount used in each field, enumerate()
function is used
'''
for index, field in enumerate(fertilizer_amount, start=1):
    print(f"N, P, K in Field {index}: {field}")

'''
To display particular nutrient amount used in each field,
enumerate() function is used in nested for loop
'''
for field_index, row in enumerate(fertilizer_amount, start=1):
     print(f"Fertilizer amounts in Field {field_index}:")
     for nutrient_index, value in enumerate(row, start=1):
         print(f" Nutrient {nutrient_index}: {value} kg")
```

Output of Listing 7.4:

The self-explanatory output of Listing 7.4 is shown below.

```
N, P, K in Field 1: [40 20 30]
N, P, K in Field 2: [35 25 32]
N, P, K in Field 3: [38 22 34]

Fertilizer amounts in Field 1:
  Nutrient 1: 40 kg
  Nutrient 2: 20 kg
  Nutrient 3: 30 kg
Fertilizer amounts in Field 2:
  Nutrient 1: 35 kg
  Nutrient 2: 25 kg
  Nutrient 3: 32 kg
Fertilizer amounts in Field 3:
  Nutrient 1: 38 kg
  Nutrient 2: 22 kg
  Nutrient 3: 34 kg
```

7.3.2.2 Arithmetic Operations on NumPy Arrays

Arithmetic operations can be performed very easily and efficiently on NumPy arrays. It is explained with the help of examples in Listing 7.5, where it is shown that applying the increase in the fertilizer amount to all fields and converting rainfall data from mm to cm is very easy using NumPy arrays.

Listing 7.5	Example of performing arithmetic operations on NumPy arrays

```
'''
NumPy Matrix containing data of Fertilizer use (N, P, K in Kgs)
across 3 fields
'''
fertilizer_amount = np.array([
    [40, 20, 30],   # Field 1
    [35, 25, 32],   # Field 2
    [38, 22, 34]    # Field 3
])

# Increase all fertilizer (or nutrient) dose by 5 units
fertilizer_amount = fertilizer_amount + 5

for field_index, row in enumerate(fertilizer_amount, start=1):
    print(f"Fertilizer amounts in Field {field_index}:")
    for nutrient_index, value in enumerate(row, start=1):
        print(f" Nutrient {nutrient_index}: {value} kg")

# Matrix containing 3 days rainfall data (in mm) on 3 Farms
rainfall_data_mm = np.array([
    [109, 122, 101],   # 3 days rainfall (in mm) on Farm 1
    [111, 131, 107],   # 3 days rainfall (in mm) on Farm 2
    [153, 142, 118]    # 3 days rainfall (in mm) on Farm 3
])

# To convert rainfall from mm to cm
rainfall_data_cm = rainfall_data_mm / 10

for index, field_index in enumerate(rainfall_data_cm, start = 1):
    print(f"Rainfall on Farm {index}: in cm is {field_index}")
```

Output of Listing 7.5:

The self-explanatory output of Listing 7.5 is shown below.

```
Fertilizer amounts in Field 1:
   Nutrient 1: 45 kg
   Nutrient 2: 25 kg
   Nutrient 3: 35 kg
Fertilizer amounts in Field 2:
   Nutrient 1: 40 kg
   Nutrient 2: 30 kg
   Nutrient 3: 37 kg
Fertilizer amounts in Field 3:
   Nutrient 1: 43 kg
   Nutrient 2: 27 kg
   Nutrient 3: 39 kg

Rainfall on Farm 1: in cm is [10.9 12.2 10.1]
Rainfall on Farm 2: in cm is [11.1 13.1 10.7]
Rainfall on Farm 3: in cm is [15.3 14.2 11.8]
```

7.3.2.3 Comparison Operations on NumPy Arrays

In NumPy, element-wise comparisons can be performed by evaluating each element in an array against either a fixed scalar value or corresponding elements (at the same positions in another array). This operation (using comparison operators such as <, >, <=, >=, ==, !=) returns a Boolean array that marks "True" wherever the specified condition holds and "False" where it does not. For example, identifying underperforming crops at different farms and the comparison of yields at multiple farms is possible through element-wise comparisons, as illustrated in Listing 7.6.

Listing 7.6	Comparison of crop yields at different agricultural farms

```
import numpy as np

'''
Crop yield (tons/hectare) for 3 crops across 3 regions of
farm1/farm2
Rows representing region → [Region1, Region2, Region3]
Columns representing crops → [Rice, Maize, Wheat]
'''
farm1_yield = np.array([
    [3.5, 4.2, 3.8],  # Region 1: Rice, Maize, Wheat
    [2.9, 3.7, 4.1],  # Region 2: Rice, Maize, Wheat
    [3.2, 4.0, 3.9]   # Region 3: Rice, Maize, Wheat
])

farm2_yield = np.array([
    [2.5, 3.5, 4.4],  # Region 1: Rice, Maize, Wheat
    [3.3, 4.4, 5.4],  # Region 2: Rice, Maize, Wheat
    [4.6, 2.1, 1.9]   # Region 3: Rice, Maize, Wheat
])

low_yield_threshold = 3.5

# Farm1/Farm2 yield's comparison with threshold value
farm1_low_yields = farm1_yield < low_yield_threshold
farm2_low_yields = farm2_yield < low_yield_threshold

# Counting total low yield values at both farms
print("Low Yields of Farm1: \n",farm1_low_yields)
print("Farm1: Total Low Yield Crops =", np.sum(farm1_low_yields))
print("Low Yields of Farm2: \n", farm2_low_yields)
print("Farm2: Total Low Yield Crops =", np.sum(farm2_low_yields))

# Comparison of two farm yields
yield_comparison = farm1_yield >= farm2_yield
print("Yield Comparsion of Farms: \n", yield_comparison)
```

```
# Comparison of Farms'yield values
for row_index, row in enumerate(yield_comparison, start=1):
    for col_index, value in enumerate(row, start=1):
        if(value == True):
            print(f"Yield at Farm1 at [{row_index}][{col_index}]
is Higher than Farm 2 Yield at [{row_index}][{col_index}]")
```

Output of Listing 7.6:

The self-explanatory output of Listing 7.6 is shown below.

```
Low yields of farm 1:
  [[False False False]
  [True False False]
  [True False False]]
Farm1: Total Low Yield Crops = 2
Low Yields of Farm2:
  [[True False False]
  [True False False]
  [False True True]]
Farm2: Total Low Yield Crops = 4
Yield comparsion of Farms:
  [[True True False]
  [False False False]
  [False True True]]
Yield at Farm 1 at [1][1] is Higher than Farm 2 Yield at [1][1]
Yield at Farm 1 at [1][2] is Higher than Farm 2 Yield at [1][2]
Yield at Farm 1 at [3][2] is Higher than Farm 2 Yield at [3][2]
Yield at Farm 1 at [3][3] is Higher than Farm 2 Yield at [3][3]
```

7.3.2.4 Indexing and Slicing

Accessing or modifying individual data points in a NumPy array is straightforward by using indexing and slicing techniques. (These techniques work similarly to the indexing and slicing approaches introduced in Chap. 5.) For example, accessing/updating a specific crop yield at different farms, a specific farm yield for different crops, or a specific crop yield in a specific field is possible through indexing and slicing, as illustrated in Listing 7.7.

Listing 7.7	Accessing and updating crop yields at agricultural farms

```
import numpy as np
'''
Crop yield data (in tons/hectare) for 3 crops across 4 fields
Rows representing Crops → [Wheat, Rice, Maize]
Columns representing Fields → [Field1, Field2, Field3, Field4]
'''
crop_yield = np.array([
    [4.2, 5.3, 3.6, 2.9],    # Wheat
    [3.7, 4.7, 4.1, 3.7],    # Rice
    [4.3, 4.6, 4.9, 4.6]     # Maize
])
print("Original Crops Yields:\n", crop_yield)
# Display yield of Corn in Field 3 (3rd row, 3rd column)
maize_field3_yield = crop_yield[2, 2]
print(f"Field 3 Maize yield: {maize_field3_yield} tons/ha")
# Display wheat yield across all fields
wheat_yield = crop_yield[0, :]
print(f"Wheat yields across all fields: {wheat_yield}")
# Display yields of field 2 for all crops
field2_yield = crop_yield[:, 1]
print(f"Wheat yields across all fields: {field2_yield}")
# Error correction: update Rice yield in Field 4 to 4.6
crop_yield[1][3] = 4.6
print("Updated Crops Yields:\n", crop_yield)
```

Output of Listing 7.7:

The self-explanatory output of Listing 7.7 is shown below.

```
Original crops yields:
 [[4.2 5.3 3.6 2.9]
 [3.7 4.7 4.1 3.7]
 [4.3 4.6 4.9 4.6]]
Field 3 Maize yield: 4.9 tons/ha
Wheat yields across all fields: [4.2 5.3 3.6 2.9]
Wheat yields across all fields: [5.3 4.7 4.6]
Updated crops yields:
 [[4.2 5.3 3.6 2.9]
 [3.7 4.7 4.1 4.6]
 [4.3 4.6 4.9 4.6]]
```

7.3.2.5 Boolean Indexing and Aggregation Functions

In NumPy, Boolean indexing allows us to filter elements from an array based on a specific condition. Moreover, aggregation functions can be applied to analyze yield performance. The usage of Boolean indexing and aggregation functions has been described in Listing 7.8.

Listing 7.8 Example: *Boolean Indexing* and *Aggregation Functions* with NumPy arrays

```
import numpy as np

'''
Crop yield data (in tons/hectare) for 3 crops across 4 fields
Rows representing Crops → [Wheat, Rice, Maize]
Columns representing Fields → [Field1, Field2, Field3, Field4]
'''

crop_yield = np.array([
    [4.2, 5.3, 3.6, 2.9],    # Wheat
    [3.7, 4.7, 4.1, 3.7],    # Rice
    [4.3, 4.6, 4.9, 4.6]     # Maize
])

print("Original Crops Yields:\n", crop_yield)

# Creating a Boolean mask based on some condition
bool_mask = crop_yield > 4.5

# Display the yield of Corn in Field 3 (3rd row, 3rd column)
high_yield = crop_yield[bool_mask]

print("Higher yield crop instances:", high_yield)

# Aggregation function to get crop yield summary
print("Average Wheat crop yield:", np.mean(crop_yield[0,:]))
print("Maximum Rice crop yield:", np.max(crop_yield[1,:]))
print("Minimum crop yield of Field 4:", np.min(crop_yield[:,3]))
print("Yields' Stand. Deviation:", round(np.std(crop_yield), 3))
```

Output of Listing 7.8:

The self-explanatory output of Listing 7.8 is shown below.

```
Original Crops Yields:
 [[4.2 5.3 3.6 2.9]
 [3.7 4.7 4.1 3.7]
 [4.3 4.6 4.9 4.6]]
Higher yield crop instances: [5.3 4.7 4.6 4.9 4.6]
Average Wheat crop yield: 4.0
Maximum Rice crop yield: 4.7
Minimum crop yield of Field 4: 2.9
Yields' Stand. Deviation: 0.635
```

7.3.2.6 NumPy Array Broadcasting

NumPy broadcasting makes it easy to do arithmetic operations on arrays that do not have the same shape. When two arrays with different shapes are used in an arithmetic operation, NumPy first checks if these shapes are compatible to perform the arithmetic operation. If it is possible, NumPy stretches the smaller array across the larger one and lets the operation happen element by element. The basic rules for broadcasting are summarized as follows:

- Two dimensions are considered a match if:
  - They are the same size, or
  - One of them is 1. If the shapes follow these rules, NumPy automatically adjusts the smaller array, so the operation can happen across both arrays smoothly.
- If the arrays have different dimensions, the shape of the smaller array is padded with ones on the left. This concept can be better understood with a simple example. Suppose you have a one-dimensional array D1 with three elements: [x, y, z], and a two-dimensional array D2 with the structure [[a], [b], [c]], where a, b, c, x, y, and z are integer values. When you attempt to add these two arrays, NumPy automatically adjusts the shape of the 1D array to match the 2D array. In this case, it broadcasts the 1D array (shape (3,)) horizontally across the rows of the 2D array (shape (3,1)). Then, it performs element-wise (row by row) addition between the arrays. A visual illustration of this NumPy broadcasting example is shown in Table 7.1.

The illustration, using specific numerical values for arrays D1 and D2, is presented in Table 7.2.

Listing 7.9 demonstrates how broadcasting works using a practical example relevant to real-time sensor-based smart irrigation systems. In such systems, a single scalar value (representing daily rainfall contribution) is broadcast across an array that holds the irrigation requirements in the absence of rain. This helps in automatically adjusting and reducing the overall irrigation needs based on the rainfall received.

Table 7.1 Visual illustration of Numpy broadcasting

Index	D2 (3×1)	Broadcasted D1 (3,)	Result
Row 0	[a]	[x, y, z]	[a+x, a+y, a+z]
Row 1	[b]	[x, y, z]	[b+x, b+y, b+z]
Row 2	[c]	[x, y, z]	[c+x, c+y, c+z]

Table 7.2 Visual illustration of Numpy broadcasting with numerical values

Index	a (3×1)	Broadcasted b (3,)	Result (a + b)
Row 0	[1]	[1, 2, 3] → [1+1, 1+2, 1+3]	[2, 3, 4]
Row 1	[2]	[1, 2, 3] → [2+1, 2+2, 2+3]	[3, 4, 5]
Row 2	[3]	[1, 2, 3] → [3+1, 3+2, 3+3]	[4, 5, 6]

Listing 7.9	Example of *Array Broadcasting* a scalar value

```
import numpy as np

# Daily irrigation need (mm) for 7 days
irrigation_schedule = np.array([15, 13, 11, 9, 10, 11, 13])
print("Daily Irrigation Schedule:", irrigation_schedule)

# Daily rainfall is 3 mm across all fields
rainfall = 3 # scalar value

# Use broadcasting to reduce irrigation by rainfall amount
adjusted_irrigation = irrigation_schedule - rainfall

print("Adjusted Daily Irrigation (mm):", adjusted_irrigation)
```

Output of Listing 7.9:

The self-explanatory output of Listing 7.9 is shown below.

Daily Irrigation Schedule: [15 13 11 9 10 11 13]
Adjusted Daily Irrigation (mm): [12 10 8 6 7 8 10]

Listing 7.10 provides another example of NumPy array broadcasting to adjust the fertilizer doses in different fields (while considering different environmental factors, i.e., soil type, crop type, or moisture levels).

Listing 7.10	Example of *Array Broadcasting* of a 1D NumPy array

```
import numpy as np

'''
Fertilizer doses (kg/hectare) for 4 crop types

Creation of 1D NumPy array named base_dose holding the original
fertilizer doses (kg/hectare) for 4 crop types (Wheat, Maize,
Oat, Rice)
'''
base_dose = np.array([45, 70, 55, 60])
print("Original Recommended Fertilizer Dose:", base_dose)
```

```
'''
Creation of 2D NumPy array called scaling_factors with 3 rows and
1 column. Considering different conditions i.e., soil type, crop
type, or moisture levels; the scaling factors represent following
adjustments to original fertilizer dose.

For example:
1.0 means no change in original fertilizer dose.
1.2 means a 20% increase in original fertilizer dose.
0.8 means a 20% decrease in original fertilizer dose.
'''
scaling_factors = np.array([[1.0], [1.2], [0.8]])

# Broadcasted multiplication to get adjusted doses for all fields
adjusted_doses = scaling_factors * base_dose

print("Adjust Fertilizer Dose: \n", adjusted_doses)
```

Output of Listing 7.10:

Original recommended fertilizer dose: [45 70 55 60]
Adjust fertilizer dose:
[[45. 70. 55. 60.]
[54. 84. 66. 72.]
[36. 56. 44. 48.]]

Explanation

The line adjusted_doses = scaling_factors * base_dose is the key line where the broadcasting of 1D array actually done. *base_dose* is a one-dimensional array consisting of 4 elements and has shape (4,). On the other hand, the scaling_factors is a two-dimensional array has shape (3, 1). The multiplication of these Numpy arrays produces a new array of shape (3, 4) that is due to the broadcasting of a **one-dimensional** array across each row of the two-dimensional array. The output represents the results of element-wise multiplication of the scaling factor array values with the base_dose array values, as shown below.

- Row 1: 1.0 * [45, 70, 55, 60]
- Row 2: 1.2 * [45, 70, 55, 60]
- Row 3: 0.8 * [45, 70, 55, 60]

7.3.2.7 Reshaping and Resizing NumPy Arrays

The reshaping and resizing are important features of NumPy and allow programmers to change the structure of arrays. Reshaping alters the layout or dimensions of an array but keeps the total number of elements the same. For example, a one-dimensional array containing six values can be reorganized into a two-dimensional array with two rows and three columns as shown below.

1D Array = [10, 20, 30, 40, 50, 60]

2D Array after reshaping → [[10, 20, 30] [40, 50, 60]]

Resizing, in contrast, modifies both the shape and the number of elements in an array. If the new size is larger, additional values are added (usually with defaults such as zeros or repeated original array values to fill the new array), and if smaller, excess values are removed. For example, a one-dimensional array containing three values can be resized into a two-dimensional array of two rows and three columns, or a two-dimensional array with two rows and two columns can be resized to a one-dimensional array with one row and two columns, as shown below.

<u>Increasing the size of array</u>

1D Array = [10, 20, 30]
2D Array after reshaping → [[10, 20, 30] [10, 20, 30]] or [[10, 20, 30] [0, 0, 0]]

<u>Decreasing the size of array</u>

2D Array = [[10, 20] [30, 40]]
2D Array after reshaping → [[10, 20]]

It is important to remember that reshaping is mainly used to reorganize existing data without loss, and resizing adapts arrays to accommodate different data sizes. Both are essential when dealing with structured information, especially in the field of agriculture, where sensor or yield data may vary in structure or volume. Listings 7.11 and 7.12 demonstrate the use of reshaping and resizing of numpy arrays in the agricultural contexts.

Listing 7.11	Example of *reshaping* 1D NumPy array data into 2D NumPy array

```
'''
Soil moisture sensor collection after every hour from 3 different
fields for 2 days. So, there are (24 hours × 2 = 48 readings per
field) and in total 144 readings for 3 fields that are stored in
1D-Array.

It is good to reshape this 1D array into a 2D with 3 rows and 48
columns (each row representing data collected in one field.
'''

import numpy as np

# Creation of assumed moisture data for 3 fields over 2 days
# (48 readings per field and in total 144 readings)
moisture_data = np.arange(144)

# Reshaping to a 3 x 48 array (3 rows: fields, 48 columns)
reshaped_data = moisture_data.reshape(3, 48)
print("Moisture Data: 3 Fields for 2 days: \n", reshaped_data)
```

Output of Listing 7.11:

Moisture Data: 3 Fields for 2 days:
[[0 1 2 3 4 5 6 7 8 9 10 11 12 13 14 15 16 17
18 19 20 21 22 23 24 25 26 27 28 29 30 31 32 33 34 35
36 37 38 39 40 41 42 43 44 45 46 47]
[48 49 50 51 52 53 54 55 56 57 58 59 60 61 62 63 64 65
66 67 68 69 70 71 72 73 74 75 76 77 78 79 80 81 82 83
84 85 86 87 88 89 90 91 92 93 94 95]
[96 97 98 99 100 101 102 103 104 105 106 107 108 109 110 111 112 113
114 115 116 117 118 119 120 121 122 123 124 125 126 127 128 129 130 131
132 133 134 135 136 137 138 139 140 141 142 143]]

Listing 7.12	Example of *resizing* 1D NumPy array into a 2D NumPy array

```
'''
Crop health record for a single season is stored in 1D array.
Need to update the structure for 3 seasons (in 2D array).
'''

# Crop health scores for 6 crop plots (1 season)
crop_health_scores = np.array([0.7, 0.6, 0.13, 0.45, 0.51, 0.37])

# Resize to hold scores for 3 seasons (will add zeros if needed)
crop_health_scores.resize((3, 6)) # 3 seasons, 6 plots
print("Crop Health Scores for 3 Seasons:\n", crop_health_scores)
```

Output of Listing 7.12:

Crop health scores for 3 seasons:
[[0.7 0.6 0.13 0.45 0.51 0.37]
[0. 0. 0. 0. 0. 0.]
[0. 0. 0. 0. 0. 0.]]

7.3.2.8 Use of *loadtxt()* and *genfromtxt()* Methods of NumPy Package

Two NumPy methods, i.e., *loadtxt()* and *genfromtxt()* allow importing of data (stored in CSV (Comma Separated Values) or Excel files). The *loadtxt()* is a simple method that does not allow the filling of missing data, but *genfromtxt*() allows it. Listings 7.13 and 7.14 describe the use of these functions in agricultural contexts.

For Listing 7.13, consider the following data stored in the ***cow_data.csv*** file.

cow_data.csv

```
CowID,Weight(kg),MilkYield(L),HealthScore
1,460,23,9
2,530,13,8
3,515,14,6
4,490,19,7
```

Listing 7.13	Example explaining the use of *loadtxt()* method of NumPy

```
import numpy as np
'''
- Load numeric data stored in cow_data.csv file
- Excluding header using skiprows =1
- delimeter refers to the character used to separate data values
'''
data = np.loadtxt('cow_data.csv', delimiter=',', skiprows=1)

# Filter cows with milk yield less than 15 L
low_health_cows = data[data[:, 2] < 15]

# Display cow records with low milk yield
print("Cows with milk yield below 15 L:\n", low_health_cows)
```

Output of Listing 7.13:

Cows with milk yield below 15 L:
[[2. 530. 13. 8.]
[3. 515. 14. 6.]]

For Listing 7.14, consider the following data stored in the ***tree_species_data.csv*** file.

tree_species_data.csv

```
Species,Carbon_Storage,Height,Age
Oak,23.5,29.6,70
Pine,18.2,,65
Maple,16.6,13,43
Birch,19.8,19,
```

Listing 7.14	Example explaining the use of *genfromtxt()* method of NumPy

```
import numpy as np
'''
- Load CSV data
- Skip header row
- Read only numeric columns
'''
tree_data = np.genfromtxt('tree_species_data.csv', delimiter=',',
            skip_header=1, usecols=(1, 2, 3), filling_values = 0)
print("Tree Data with filled values: \n", tree_data)

# Calculate averages
avg_carbon = np.mean(tree_data[:, 0])# carbon storage (col. 1)
avg_height = np.mean(tree_data[:, 1]) # height (col. 2)
avg_age = np.mean(tree_data[:, 2]) # age (col. 3)

# Displaying calculated averages
print(f"Average Carbon Storage (tons): {avg_carbon:.2f}")
print(f"Average Tree Height (meter): {avg_height:.2f}")
print(f"Average Age (years): {avg_age:.2f}")
```

Output of Listing 7.14:

```
Tree Data with filled values:
[[23.5 29.6 70.]
[18.2 0. 65.]
[16.6 13. 43.]
[19.8 19. 0.]]
Average carbon storage (tons): 19.53
Average tree height (meter): 15.40
Average age (years): 44.50
```

7.3.2.9 Case Study 7.1: Selection of High-Yielding Varieties Using Trait Data

A plant breeder is interested in analyzing various varieties of a crop on the basis of genotype data of that crop to identify the ideal varieties for future breeding. The data collected from multiple field trials includes measurements such as:

- Plant height (cm)
- Number of grains per spike
- 1000-grain weight (g)
- Yield per hectare (tons)

and stored in a file named ***crop_varieties_data.csv*** (as shown in the sample given below).

crop_varieties_data.csv

```
CropVariety,PlantHeight,GrainsPerSpike,GrainWeight,Yield
V1,79,45,35,3.2
V2,95,50,38,3.5
V3,86,42,33,3.0
V4,93,48,36,3.6
V5,91,40,32,2.8
V6,89,46,37,3.4
```

Write a Python program for that plant breeder to calculate averages of varieties' traits (i.e., plant height, grains per spike, grain weight, yield), finding underperforming crop varieties, and identify ideal crop varieties for future breeding. Moreover, the plant breeder is also interested in ranking varieties based on yield.

Listing 7.15	Solution of case study 7.1

```
import numpy as np

# Loading crop variety names separately (in column 0)
variety_name = np.genfromtxt("crop_varieties_data.csv",
delimiter=",", skip_header=1, usecols=(0), dtype=str)

# Loading data excluding crop varieities (col. 1 to col. 4)
data = np.genfromtxt("crop_varieties_data.csv", delimiter=",",
                               skip_header=1, usecols=(1, 2, 3, 4))

# Column indices (declared as constants) for readability
PLANT_HEIGHT = 0
GRAINS_PER_SPIKE = 1
GRAIN_WEIGHT = 2
YIELD = 3

print("==================================================")

# Calculate averages of crop varieties' traits
averages = data.mean(axis=0)
print("Averages of Crop Varieties' Trait Values:")
print(f"Plant Height: {averages[PLANT_HEIGHT]:.2f} cm")
print(f"Grains/Spike: {averages[GRAINS_PER_SPIKE]:.2f}")
print(f"Grain Weight: {averages[GRAIN_WEIGHT]:.2f} g")
print(f"Crop Yield: {averages[YIELD]:.2f} t/ha")
print("==================================================")

'''
Selecting top-performing crop varieties by filtering varieties
above average in yield
'''
```

```
selected = (data[:, YIELD] > averages[YIELD])
top_yield_varieties = variety_name[selected]
print("Ideal Crop Varieties (> avg. in yield) for Breeding :")
print(top_yield_varieties)
print("================================================")

'''
To do a ranking of crop varieties by yield, use the argsort()
method of numpy instead of using the sort() method because the
sort() method returns the sorted values, and argsort() returns
the indices, which is helpful to sort array values.

[::-1] in the below statement represents [start:stop:step] that
is explained below:
start = not specified → start from the end
stop = not specified → go all the way to the beginning
step = -1 → move backwards, one step at a time to reverse into
descending order.
'''

sorted_indices = np.argsort(data[:, YIELD])[::-1]
print("Ranking of Crop Varieties by Yield:")
for index in sorted_indices:
    print(f"{variety_name[index]} -> Yield: {data[index,
YIELD]} t/ha")

print("================================================")
```

Output of Listing 7.15:

The self-explanatory output of Listing 7.15 is shown below.

```
================================================
Averages of crop varieties' trait values:
Plant height: 88.83 cm
Grains/spike: 45.17
Grain weight: 35.17 g
Crop yield: 3.25 t/ha
================================================
Ideal crop varieties (> avg. in yield) for breeding:
['V2' 'V4' 'V6']
================================================
Ranking of crop varieties by yield:
V4 -> Yield: 3.6 t/ha
V2 -> Yield: 3.5 t/ha
V6 -> Yield: 3.4 t/ha
V1 -> Yield: 3.2 t/ha
V3 -> Yield: 3.0 t/ha
V5 -> Yield: 2.8 t/ha
================================================
```

7.4 Pandas

Pandas is an open-source Python package that is built on top of NumPy and is widely used for data analysis, particularly with structured data such as tables and spreadsheets. It simplifies many preprocessing data-related tasks that are often repetitive and time-consuming (i.e., data loading, cleaning, filling missing values, merging or joining datasets) before performing statistical analysis.

7.4.1 Types of Pandas Data Structures

Mainly, there are two types of data structures that are supported by Pandas, i.e., Series and DataFrame. A Series is either a standalone one-dimensional data structure or represents a single column or row of data within a DataFrame. A DataFrame is a two-dimensional data structure that is used to handle tabular data with multiple rows and columns. Agricultural data is often large and structured (e.g., crop yields from different regions or seasons, soil nutrient reports, weather data, pest monitoring records, etc.), and therefore Pandas with Series and DataFrame structures make it easy to load, organize, manage, filter, and analyze this kind of structured agricultural data. Both Series and DataFrames are foundational to data science workflows in Python and are essential tools for anyone, including agriculturists working with structured data. (*Note*: Structured data has a standardized format, typically a tabular format with rows and columns (defining clear data attributes) and is effectively processed by computing techniques for insights. Examples of structured data include data stored in relational databases, Excel files, web forms, and so on. Other than Structured data, Semi-structured data, and Unstructured data are the other two types of data. Semi-structured data is not fully structured data as it lacks completeness of information and does not exist in tabular form. However, Semi-structured datasets include metadata (with tags and markers) that is helpful for analysis. Examples of Semi-structured data are emails, zipped files, JSON (JavaScript Object Notation), CSV (Comma-Separated Values) files, and so on. The unstructured datasets have an internal structure but lack a predefined schema or format. Examples of unstructured data include audio/video files, photo files, text files, and so on.)

7.4.2 Benefits of Using Pandas for Agricultural Data

In the field of agriculture, Pandas is especially valuable for managing and analyzing large datasets such as weather records, crop yields, soil test results, pest observations, and livestock information. For instance, an agriculturist can use Pandas to explore yield trends across seasons and locations, examine the relationship between fertilizer or pesticide usage and production outcomes, or clean and organize sensor data for better decision-making. With powerful features for handling missing data, merging data from diverse sources, and conducting time-series analysis, Pandas makes the application of data science in agriculture more accessible and impactful.

Listings 7.15–7.24 include examples that demonstrate how to use Python's Pandas package and show how it can be applied to perform various types of agricultural data analysis. These examples highlight different uses of Pandas, from organizing and cleaning data to generating insights from datasets containing records of crop yields, weather conditions, soil tests, pest infestations, tracking livestock health, and so on. Each listing demonstrates how the Pandas package supports effective data handling in agriculture.

7.4.2.1 Creation and Use of Python Pandas Series

Listing 7.16 illustrates the creation of Pandas Series that are useful for allowing easy access to a single column of data and element-wise operations. Moreover, it also demonstrates the use of associated methods that can be used for statistical analysis.

Listing 7.16	Example explaining the creation of Python Pandas Series and use of a few associated methods

```
import pandas as pd

# Creation of pandas Series (data with given index names)
soil_pH = pd.Series([3.5, 9.7, 15.8, 13.3], index=['Plot1', 'Plot2',
                                                    'Plot3', 'Plot4'])

# Display pandas Series data and related index
print("Soil pH Values:\n", soil_pH)
print("==============================")

# To display pandas Series without dtype
print("Soil pH Values:\n", soil_pH.to_string(dtype=False))
print("==============================")

print("Total number of soil samples:", soil_pH.count())
print("Average pH Value of soil samples:", soil_pH.mean())
print("Maximum pH Value in soil samples:", soil_pH.max())
print("Minimum pH Value in soil samples:", soil_pH.min())
print("==============================")

print("Overall Dataset Description")
print(soil_pH.describe())
print("==============================")
```

Output of Listing 7.16:

The self-explanatory output of Listing 7.16 is shown below.

```
Soil      pH values:
Plot1       3.5
Plot2       9.7
Plot3      15.8
Plot4      13.3
dtype:    float64
=============================
Soil      pH values:
Plot1       3.5
Plot2       9.7
Plot3      15.8
Plot4      13.3
=============================
Total number of soil samples: 4
Average pH value of soil samples: 10.575
Maximum pH value in soil samples: 15.8
Minimum pH value in soil samples: 3.5
=============================
Overall dataset description
count     4.000000
mean     10.575000
std       5.340022
min       3.500000
25%       8.150000
50%      11.500000
75%      13.925000
max      15.800000
dtype: float64
=============================
```

7.4.2.2 Creation and Use of Python Pandas DataFrame

Listing 7.17 illustrates the creation of Pandas DataFrames using different Python iterables, i.e., List, Dictionary, and Numpy array, that ultimately are useful for providing powerful functionality for handling, analyzing, and visualizing large datasets. Using Pandas DataFrames, users can filter data, perform group operations, merge datasets, and apply complex transformations efficiently.

Listing 7.17	Example explaining the creation of Python Pandas DataFrame and the use of a few associated methods

```
import pandas as pd

# Create a list of rows
list_data = [['Cow1',15,'Yes'],['Cow2',28,'No'],['Cow3',22,'Yes']]
# Define column labels
columns = ["AnimalID", "Age", "Vaccinated"]

# Creation of Pandas DataFrame (using List Data)
livestock_df = pd.DataFrame(list_data, columns = columns)
print("Total number of data entries:")
print(livestock_df.count())
print("============================")
print("Maximum Values in Dataset:")
print(livestock_df.max())
print("============================")
print("Minimum Values in Dataset:")
print(livestock_df.min())
print("============================")
print("Overall Dataset Description")
print(livestock_df.describe())
print("============================")

# Data stored in Python Dictionary
dic_data = {
    'Crop': ['Maize', 'Wheat', 'Rice'],
    'ExpectedYield': [3.1, 2.8, 2.4],
    'ActualYield': [1.1, 2.1, 1.4],
    'Irrigation': ['Drip', 'Sprinkler', 'Flood']
}

# Creation of Pandas DataFrame (using Dictionary Data)
crop_df = pd.DataFrame(dic_data)
print("Total number of data entries:")
print(crop_df.count())
```

```
print("============================")
print("Maximum Values in Dataset:")
print(crop_df.max())
print("============================")
print("Minimum Values in Dataset:")
print(crop_df.min())
print("============================")
print("Overall Dataset Description")
print(crop_df.describe())
print("============================")

import numpy as np
import pandas as pd
# Create a NumPy array
plant_data = np.array([["G1", 45, 35], ["G2", 42, 33], ["G3",
43, 37]])

# Define column labels
columns = ["Genotype", "PlantHeight", "GrainsPerSpike"]

# Create a DataFrame from the NumPy array
plant_df = pd.DataFrame(plant_data, columns = columns)

print("Total number of data entries:")
print(plant_df.count())
print("============================")
print("Maximum Values in Dataset:")
print(plant_df.max())
print("============================")
print("Minimum Values in Dataset:")
print(plant_df.min())
print("============================")
print("Overall Dataset Description")
print(plant_df.describe())
print("============================")
```

Output of Listing 7.17:

```
Total number of data entries:
AnimalID          3
Age               3
Vaccinated        3
dtype: int64
==============================
Maximum Values in the Dataset:
AnimalID Cow3
Age               28
Vaccinated        Yes
dtype: object
==============================
Minimum Values in Dataset:
AnimalID Cow1
Age               15
Vaccinated        No
dtype: object
==============================
Overall Dataset Description
Age
count          3.000000
mean          21.666667
std            6.506407
min           15.000000
25%           18.500000
50%           22.000000
75%           25.000000
max           28.000000
==============================
Total number of data entries:
Crop              3
ExpectedYield     3
ActualYield       3
Irrigation        3
dtype: int64
==============================
Maximum Values in Dataset:
Crop Wheat
ExpectedYield     3.1
ActualYield       2.1
Irrigation Sprinkler
dtype: object
==============================
Minimum values in dataset:
Crop Maize
ExpectedYield     2.4
ActualYield       1.1
Irrigation Drip
dtype: object
==============================
Overall Dataset Description
ExpectedYield     ActualYield
count   3.000000   3.000000
mean    2.766667   1.533333
```

```
std         0.351188       0.513160
min         2.400000       1.100000
25%         2.600000       1.250000
50%         2.800000       1.400000
75%         2.950000       1.750000
max         3.100000       2.100000
=============================
Total number of data entries:
Genotype          3
PlantHeight       3
GrainsPerSpike    3
dtype: int64
=============================
Maximum values in dataset:
Genotype          G3
PlantHeight       45
GrainsPerSpike    37
dtype: object
=============================
Minimum values in dataset:
Genotype          G1
PlantHeight       42
GrainsPerSpike    33
dtype: object
=============================
Overall dataset description
Genotype PlantHeight GrainsPerSpike
count          3 3 3
unique         3 3 3
top         G1 45 35
freq           1 1 1
=============================
```

Explanation:

This example demonstrates that a Pandas DataFrame can be created using Python lists, dictionaries, or NumPy arrays. Unlike basic data structures, a DataFrame presents data in a well-organized, tabular format with support for labeled indexing, making it easier to interpret and analyze. Although much of the output is intuitive, certain parts needs further explanation. For instance, you may notice the following lines in the output summary:

```
25% 18.500000
50% 22.000000
75% 25.000000
```

and the values labeled here as 25%, 50%, and 75% represent percentiles. Specifically, the 25% percentile indicates that 25% of the data points are less than or equal to approximately 18.5. The 50% percentile, also known as the median, shows the midpoint value of the dataset, which in this case is 22.0. The 75% percentile tells us that 75% of the data values fall below or are equal to 25.0.

Additionally, you may notice a line in the summary output that reads:

```
freq 1 1 1
```

This means that the most frequently occurring value (displayed as "top" in the same output) appears only once in each respective column. Thus, the freq value represents the count of the top (most frequent) item within each column.

7.4.2.3 Common Operations on Python Pandas DataFrames

Pandas DataFrames provide a wide range of operations that allow users to efficiently manage and analyze structured data. For example, you can

- Retrieve all or selected records from specific rows and columns
- Review and understand the overall structure and contents of the given data
- Handling missing values
- Removing duplicate data entries
- Look up information about specific data item(s) based on certain criteria
- Saving DataFrame to CSV file

Consider the following case study to understand how these common operations can be performed using Python Pandas DataFrames.

Case Study 7.2: Tracking animal health information

A veterinary doctor wants to analyze and explore the animal health data (shown below) stored in a CSV file named *veterinary_records.csv* in the following ways:

- Retrieve all or selected records from specific rows and columns
- Review and understand the overall structure and contents of the veterinary data
- Check the vaccination status of each animal
- Look up information about a specific animal based on certain criteria
- Identify animals that have not been vaccinated
- Find unvaccinated animals with a body temperature above 39°C
- Handling missing values
- Removing duplicate data entries
- Saving DataFrame to CSV file

veterinary_records.csv

```
AnimalID,Breed,Temperature,Vaccinated,Notes
Cow1,Holstein,39.2,Yes,Normal
Cow2,Jersey,40.1,No,Fever
Cow3,Sahiwal,38.7,Yes,Slight cough
Cow4,Holstein,40.3,No,High fever
Cow5,Dhanni,41.1,No,High fever
Cow6,Tharparkar,38.7,No,Mastitis
Cow7,Tuli,38.7,No,BVD
Cow7,Tuli,38.7,No,BVD
Cow8,Taurus,,Yes,BVD
Cow9,Brahman,38.2,,Mastitis
Cow10,Deoni,38.0,No,
```

The implementation of this case study in Listing 7.18 illustrates not only how to create DataFrames by importing data from *veterinary_records.csv* file but also guides you through essential data handling tasks such as:

- Inspecting and summarizing large datasets for quick understanding
- Extracting specific and relevant information
- Cleaning and preparing data subsets for further analysis or reporting, which includes:
  - Managing missing or incomplete values
  - Removing duplicate entries and filtering rows based on specific conditions
- Saving DataFrame to CSV file.

Listing 7.18	Solution of Case Study 7.2

```
import pandas as pd

# Read veterinary data from CSV file
vet_df = pd.read_csv('veterinary_records.csv')

# Display read veterinary data from CSV file
print("========================================")
print("Overall Veterinary Record as DataFrame")
print("========================================")
print(vet_df)

# Getting data in rows using head() and tail() methods
print("================================================")
print("Use of head() and tail() Methods of DataFrames")
print("================================================")
'''
Starting from stop (including column names), head() method
returns a specified number of rows. By default (if no number
specified), it returns the first 5 rows.
'''

print("Default (Five) Rows of Veterinary Data")
print(vet_df.head())
print("First Three Rows of Veterinary Data")
print(vet_df.head(3))

print("Default Last (Five) Rows of Veterinary Data")

'''
Starting from bottom (but including column names), tail() method
returns a specified number of rows. By default (if no number
specified), it returns the Last 5 rows.
'''
```

```
print(vet_df.tail())
print("Last Three Rows of Veterinary Data")
print(vet_df.tail(3))

# Displaying information of DataFrame structure
print("====================================")
print("Use of info() Methods see DataFrame Structure")
print("====================================")
print(vet_df.info())

# Displaying column data using get() method or square brackets
print("========================================================")
print("Access Data using get() Method & Square Brackets []")
print("========================================================")

# 1. Using get() method of DataFrame
print("All Animal IDs")
print(vet_df.get("AnimalID"))

# 2. Using cloumn name in square bracket
print("Animals' Vaccination Status")
print(vet_df["Vaccinated"])

# Display row(s), column(s) & cells using loc[], iloc[]
print("====================================")
print("Use of loc[], iloc[] to Access Specific row(s), column(s)")
print("====================================")
'''
1. Use of loc[]
- loc[] is used with label-level indexing
  (means to select specific data subset using row/column labels)
- iloc [] is used with position-level indexing
  (means to select specific data subset using indices)

The general syntax of loc[] is:

    df.loc[row_label, column_label]

Important Note:

By default, DataFrames assign row labels automatically e.g. it
assigned 1, 2, 3, ... as row labels to CSV data. However, to set
first column as row label you can use the index_col parameter when
reading data from a CSV or Excel file as shown below

                    vet_df = pd.read_csv('veterinary_records.csv')

But if dataset is already loaded then you can use set_
index() method
and after that you can use these column values as row labels as
demonstrated in the code below.
'''
```

```
vet_df.set_index("AnimalID", inplace = True)
'''
Alternatively, you can also use the set_index method in the
following way.

    vet_df.set_index(vt_df.columns[0], inplace=True)
'''

# Getting all details of Cow1
print("Details of Cow1")
print(vet_df.loc["Cow1"])

# Getting details of only the 'Temperature' and 'Notes' of Cow3
print(vet_df.loc["Cow3", ['Temperature', 'Notes']])

# Getting the details of all cows with 'High fever'
print(vet_df.loc[vet_df['Notes'] == 'High fever'])

# Getting details of unvaccinated Cows
unvaccinated_cases = vet_df[(vet_df['Temperature'] > 39.5) &
                            (vet_df['Vaccinated'] == 'No')]

print("Unvaccinated Veterinary Cases")
print(unvaccinated_cases)

# Critical case (Unvaccinated cows with temp. > 39)
critical_cases = vet_df[(vet_df['Temperature'] > 39) &
                        (vet_df['Vaccinated'] == 'No')]
print("Critical Veterinary Cases")
print(critical_cases)

'''
2. Use of iloc[]
The general syntax of using iloc[] is
    df.iloc[row_index, column_index]
It selects data by integer position (row and column positions).
'''

# Getting the data in the first row
print("Data in Row 1")
print(vet_df.iloc[0])

# Getting the values at 2nd row, 3rd column
print("Cow3's Temperature")

'''
Temperature index is 1 because we already set column 0 (AnimalID)
is as index
'''
print(vet_df.iloc[2, 1])
```

```
# Getting the first three rows and first two columns data
print("Data Stored in First 3 Rows and First Two Columns")
print(vet_df.iloc[0:3, 0:2])

# Detection of missing values
print("====================================")
print("Handling Missing ")
print("====================================")

print("Displaying of Null Values as Boolean True")
print(vet_df.isnull())

# Count total missing values in each column
print("Total Null Values in each Column")
print(vet_df.isnull().sum())

'''
replacing missing values with default/calculated value (using
fillna()).Replace missing temperature with average temperature in
vet_df
'''
print("Filling of Null Values with Some Calculated Value")
avg_temp = vet_df['Temperature'].mean()
vet_df['Temperature'] = vet_df['Temperature'].fillna(avg_temp)
print("Replacing Null Values with Specific Value")

# Replace missing vaccinated information with 'Not Available'
vet_df['Vaccinated'] = vet_df['Vaccinated'].fillna('Not
Available')

# Replace missing Notes information with 'Not Known'
vet_df['Notes'] = vet_df['Notes'].fillna('Not known')
print(vet_df)

# Detection of duplicated data entries
print("Display of Duplicated Data Entries")
print(vet_df.duplicated())

# Counting of duplicated data entries
print("Total Count of Duplicated Data Entries")
print(vet_df.duplicated().sum())

# Remove duplicate data entries in vet_df
final_vet_df = vet_df.drop_duplicates()

print("Final Dataset after Removal Duplicated Records")
print(final_vet_df)

print("Saving of DataFrames in CSV File")

# Saving modified dataset to a CSV file without exporting index
final_vet_df.to_csv('modified_veterinary_records.csv', index=False)
print("Modified Dataset has been Stored Successfully")
```

Output of Listing 7.18:

The self-explanatory output of Listing 7.18 is shown below.

```
=====================================
Overall Veterinary Record as DataFrame
=====================================
   AnimalID      Breed   Temperature  vaccinated         Notes
0      Cow1   Holstein          39.2         Yes        Normal
1      Cow2     Jersey          40.1          No         Fever
2      Cow3    Sahiwal          38.7         Yes  Slight cough
3      Cow4   Holstein          40.3          No    High fever
4      Cow5     Dhanni          41.1          No    High fever
5      Cow6 Tharparkar          38.7          No      Mastitis
6      Cow7       Tuli          38.7          No           BVD
7      Cow7       Tuli          38.7          No           BVD
8      Cow8     Taurus           NaN         Yes           BVD
9      Cow9    Brahman          38.2         NaN      Mastitis
10    Cow10      Deoni          38.0          No           NaN
=====================================
Use of head() and tail() Methods of DataFrames
=====================================
Default (Five) Rows of Veterinary Data
   AnimalID      Breed   Temperature  Vaccinated         Notes
0      Cow1   Holstein          39.2         Yes        Normal
1      Cow2     Jersey          40.1          No         Fever
2      Cow3    Sahiwal          38.7         Yes  Slight cough
3      Cow4   Holstein          40.3          No    High fever
4      Cow5     Dhanni          41.1          No    High fever
First Three Rows of Veterinary Data
   AnimalID      Breed   Temperature  Vaccinated         Notes
0      Cow1   Holstein          39.2         Yes        Normal
1      Cow2     Jersey          40.1          No         Fever
2      Cow3    Sahiwal          38.7         Yes  Slight cough
Default Last (Five) Rows of Veterinary Data
   AnimalID      Breed   Temperature  Vaccinated         Notes
6      Cow7       Tuli          38.7          No           BVD
7      Cow7       Tuli          38.7          No           BVD
8      Cow8     Taurus           NaN         Yes           BVD
9      Cow9    Brahman          38.2         NaN      Mastitis
10    Cow10      Deoni          38.0          No           NaN
Last Three Rows of Veterinary Data
   AnimalID      Breed   Temperature  Vaccinated         Notes
8      Cow8     Taurus           NaN         Yes           BVD
9      Cow9    Brahman          38.2         NaN      Mastitis
10    Cow10      Deoni          38.0          No           NaN
=====================================
Use of info() Methods see DataFrame Structure
=====================================
<class 'pandas.core.frame.DataFrame'>
RangeIndex: 11 entries, 0 to 10
Data columns (total 5 columns):
```

```
#   Column        Non-NullCount    Dtype
---  ------        --------------   -----
0   AnimalID      11 non-null      object
1   Breed         11 non-null      object
2   Temperature   10 non-null      float64
3   Vaccinated    10 non-null      object
4   Notes         10 non-null      object
dtypes: float64(1), object(4)
memory usage: 572.0+ bytes
None
======================================
Access Data using get() Method & Square Brackets []
======================================
All Animal IDs
0     Cow1
1     Cow2
2     Cow3
3     Cow4
4     Cow5
5     Cow6
6     Cow7
7     Cow7
8     Cow8
9     Cow9
10    Cow10
Name: AnimalID, dtype: object
Animals' vaccination status
0     Yes
1     No
2     Yes
3     No
4     No
5     No
6     No
7     No
8     Yes
9     NaN
10    No
Name: Vaccinated, dtype: object
================================================
Use of loc[], iloc[] to Access Specific row(s), column(s) and cells
================================================
Details of Cow1
Breed          Holstein
Temperature    39.2
Vaccinated     Yes
Notes          Normal
Name: Cow1, dtype: object
Temperature    38.7
Notes          Slight cough
Name: Cow3, dtype: object
   AnimalID    Breed     Temperature   Vaccinated    Notes
   Cow4        Holstein  40.3          No            High fever
```

```
Cow5    Dhanni    41.1    No    High fever
Unvaccinated Veterinary Cases
 Breed Temperature Vaccinated Notes
AnimalID
Cow2    Jersey    40.1    No    Fever
Cow4    Holstein  40.3    No    High fever
Cow5    Dhanni    41.1    No    High fever
Critical Veterinary Cases
 Breed Temperature Vaccinated Notes
AnimalID
Cow2    Jersey    40.1    No    Fever
Cow4    Holstein  40.3    No    High fever
Cow5    Dhanni    41.1    No    High fever
Data in Row 1
Breed           Holstein
Temperature     39.2
Vaccinated      Yes
Notes           Normal
Name: Cow1, dtype: object
Cow3's Temperature
        38.7
Data Stored in First 3 Rows and First Two Columns
 Breed    Temperature
AnimalID
Cow1    Holstein  39.2
Cow2    Jersey    40.1
Cow3    Sahiwal   38.7
====================================
Handling Missing
====================================
Displaying of Null Values as Boolean True
Breed Temperature Vaccinated Notes
AnimalID
Cow1    False    False    False    False
Cow2    False    False    False    False
Cow3    False    False    False    False
Cow4    False    False    False    False
Cow5    False    False    False    False
Cow6    False    False    False    False
Cow7    False    False    False    False
Cow7    False    False    False    False
Cow8    False    True     False    False
Cow9    False    False    True     False
Cow10   False    False    False    True
Total Null Values in each Column
Breed           0
Temperature     1
Vaccinated      1
Notes           1
dtype: int64
Filling of Null Values with Some Calculated Value
Replacing Null Values with Specific Value
 Breed    Temperature    Vaccinated    Notes
```

```
AnimalID
Cow1      Holstein     39.20   Yes            Normal
Cow2      Jerse        40.10   No             Fever
Cow3      Sahiwal      38.70   Yes            Slight cough
Cow4      Holstein     40.30   No             High fever
Cow5      Dhanni       41.10   No             High fever
Cow6      Tharparkar   38.70   No             Mastitis
Cow7      Tuli         38.70   No             BVD
Cow7      Tuli         38.70   No             BVD
Cow8      Taurus       39.17   Yes            BVD
Cow9      Brahman      38.20   Not Available  Mastitis
Cow10     Deoni        38.00   No             Not known
Display of Duplicated Data Entries
AnimalID
Cow1      False
Cow2      False
Cow3      False
Cow4      False
Cow5      False
Cow6      False
Cow7      False
Cow7      True
Cow8      False
Cow9      False
Cow10     False
dtype:    bool
Total Count of Duplicated Data Entries   1
Final Dataset after Removal Duplicated Records
Breed Temperature Vaccinated Notes
AnimalID
Cow1      Holstein     39.20   Yes    Normal
Cow2      Jersey       40.10   No     Fever
Cow3      Sahiwal      38.70   Yes    Slight cough
Cow4      Holstein     40.30   No     High fever
Cow5      Dhanni       41.10   No     High fever
Cow6      Tharparkar   38.70   No     Mastitis
Cow7      Tuli         38.70   No     BVD
Cow8      Taurus       39.17   Yes    BVD
Cow9      Brahman      38.20   Not    Available Mastitis
Cow10     Deoni        38.00   No     Not known
Saving of DataFrames in CSV File
Modified Dataset has been Stored Successfully
```

7.4.2.4 Customized Filtering in Pandas DataFrames

Customized filtering with Pandas DataFrames is extremely useful in agricultural settings, allowing users to extract specific records that meet certain conditions. For instance, the example in Listing 7.19 demonstrates how filtering can be applied to retrieve targeted data based on defined criteria.

Listing 7.19	Example: *Customized Filtering* using Pandas DataFrames

```
# Customized Filtering using Pandas DataFrames
print("==============================================")
print("Customized Filtering using Pandas DataFrames")
print("==============================================")

import pandas as pd

field_dataset = {
    'Crop': ['Rice', 'Barley', 'Wheat', 'Soybean', 'Maize'],
    'Yield': [3.2, 3.1, 2.8, 3.5, 5.9], # (tons/ha)
    'Rainfall': [650, 300, 440, 490, 310], # mm
    'Soil pH': [5.5, 6.0, 5.5, 7.2, 6.3]
}

field_df = pd.DataFrame(field_dataset)
print("Original Dataset")
print(field_df)
print("=====================================")

# Selection of crops with yield greater than 3 tons/ha
high_yield_crops = field_df[field_df['Yield'] > 3]
print("Crops with higher yield")
print(high_yield_crops)
print("=====================================")

# Selection of crops grown in areas with rainfall < 450 mm
low_rainfall_area_crops = field_df[field_df['Rainfall'] < 450]
print("Crops in low rainfall area")
print(low_rainfall_area_crops)
print("=====================================")

# Selection of crops with soil pH between 6.0 and 7.0
optimal_ph_crops = field_df[(field_df['Soil pH'] >= 6.0) &
                                        (field_df['Soil pH'] <= 7.0)]

print("Crops optimal pH soil")
print(optimal_ph_crops)
print("=====================================")
```

Output of Listing 7.19:

The self-explanatory output of Listing 7.19 is shown below.

```
=====================================
Customized Filtering using Pandas DataFrames
=====================================
Original Dataset
 Crop Yield Rainfall Soil pH
0 Rice         3.2 650 5.5
1 Barley       3.1 300 6.0
2 Wheat        2.8 440 5.5
3 Soybean      3.5 490 7.2
4 Maize        5.9 310 6.3
=====================================
Crops with higher yield
 Crop Yield Rainfall Soil pH
0 Rice         3.2 650 5.5
1 Barley       3.1 300 6.0
3 Soybean      3.5 490 7.2
4 Maize        5.9 310 6.3
=====================================
Crops in low rainfall area
 Crop Yield Rainfall Soil pH
1 Barley       3.1 300 6.0
2 Wheat        2.8 440 5.5
4 Maize        5.9 310 6.3
=====================================
Crops optimal pH soil
 Crop Yield Rainfall Soil pH
1 Barley       3.1 300 6.0
4 Maize        5.9 310 6.3
=====================================
```

7.4.2.5 Sorting and Ranking Operations on Pandas DataFrames

Sorting and Ranking operations using Pandas DataFrames (especially in the context of agriculture) are sometimes helpful in analyzing and comparing agricultural data; for instance, sorting crops by yield or ranking farms based on productivity, as demonstrated in Listing 7.20.

Listing 7.20	Example: *Sorting* and *Ranking* using Pandas DataFrames

```
# Sorting and Ranking using Pandas DataFrames
print("==========================================")
print("Sorting and Ranking using Pandas DataFrames")
print("==========================================")

import pandas as pd

farm_data = {
    'Farm': ['Farm A', 'Farm B', 'Farm C', 'Farm D'],
    'Crop': ['Rice', 'Wheat', 'Maize', 'Barley'],
    'Yield (tons)': [310, 270, 195, 275],
    'Date': ['2024-09-01', '2024-09-05', '2024-08-30',
'2024-09-03']
}

farm_df = pd.DataFrame(farm_data)
print("Original Dataset")
print(farm_df)
print("====================================")

# Sorting by Yield (Ascending Order)
sorted_by_yield = farm_df.sort_values(by='Yield (tons)')
print("Dataset sorted (in ascending order) by yield")
print(sorted_by_yield)
print("====================================")

# Sorting by Yield (Descending Order)
sorted_by_yield = farm_df.sort_values(by='Yield (tons)',
ascending=False)
print("Dataset sorted (in descending order) by yield")
print(sorted_by_yield)
print("====================================")

# Sorting by Date (Most Recent First)
sorted_by_date = farm_df.sort_values(by='Date', ascending=False)
print("Dataset sorted (in descending order) by date")
print(sorted_by_date)
print("====================================")

# Assigning Rank to Farms by Yield (Higher is Better)
farm_df['Rank'] = farm_df['Yield (tons)'].rank(ascending=False)
print("Dataset ranked by yield")
print(farm_df[['Farm', 'Yield (tons)', 'Rank']])
print("====================================")
```

Output of Listing 7.20:

The self-explanatory output of Listing 7.20 is shown below.

```
=====================================
Sorting and Ranking using Pandas DataFrames
=====================================
Original Dataset
 Farm Crop Yield (tons) Date
0Farm A      Rice      310 2024-09-01
1Farm B      Wheat     270 2024-09-05
2Farm C      Maize     195 2024-08-30
3Farm D      Barley    275 2024-09-03
=====================================
Dataset sorted (in ascending order) by yield
 Farm Crop Yield (tons) Date
2Farm C      Maize     195 2024-08-30
1 Farm B     Wheat     270 2024-09-05
3 Farm D     Barley    275 2024-09-03
0 Farm A     Rice      310 2024-09-01
=====================================
Dataset sorted (in descending order) by yield
 Farm Crop Yield (tons) Date
0 Farm A     Rice      310 2024-09-01
3 Farm D     Barley    275 2024-09-03
1 Farm B     Wheat     270 2024-09-05
2 Farm C     Maize     195 2024-08-30
=====================================
Dataset sorted (in descending order) by date
 Farm Crop Yield (tons) Date
1 Farm B     Wheat     270 2024-09-05
3 Farm D     Barley    275 2024-09-03
0 Farm A     Rice      310 2024-09-01
2 Farm C     Maize     195 2024-08-30
=====================================
Dataset ranked by yield
Farm Yield (tons) Rank
0 Farm A     310 1.0
1 Farm B     270 3.0
2 Farm C     195 4.0
3 Farm D     275 2.0
=====================================
```

7.4.2.6 Merging and Joining Pandas DataFrames

In real-world scenarios, most of the time (agricultural) data comes from multiple sources, i.e., crop field records, in-field sensors, weather stations, and soil testing labs. In these situations, efficiently combining these datasets is essential for meaningful analysis and insights. Combining datasets using Pandas DataFrame is possible through merge(), concat(), and join(), as illustrated in Listing 7.21.

Listing 7.21 Example: Combining of agricultural datasets using *merge()*, *join()*, and *concat()*

```
# Combine Agri. Data using merge(),join(),and concat() methods
print("=====================================")
print(" Use of merge(),join(), and concat() Methods")
print("=====================================")

import pandas as pd

# Soil Dataset from different fields of an Agriculture Farm
soil_data = pd.DataFrame({
    'Field_ID': ["F1", "F2", "F3", "F4"],
    'Soil_pH': [5.5, 6.7, 7.2, 6.5],
    'Nitrogen': [35, 45, 49, 39]
})

print("Original soil dataset")
print(soil_data)
print("=====================================")

# Weather Dataset from different fields of an Agriculture Farm
weather_data = pd.DataFrame({
    'Field_ID': ["F1", "F2", "F3", "F5"],
    'Rainfall_mm': [110, 120, 95, 88],
    'Temperature': [24, 26, 23, 25]
})

print("Original weather dataset")
print(weather_data)
print("=====================================")

print("Using of merge() method")
print("=====================================")
'''
The merge() method combines DataFrames based on common column(s).
It can be Inner Merge, Left Merge, Right Merge, and Outer Merge.
'''

# 1. Inner Merge - Only matching Field_ID entries
#Use case: Analyze data only for fields with both soil weather
records.

merged_inner = pd.merge(soil_data, weather_data, on='Field_ID',
                                                      how='inner')

print("Dataset after Inner Merge")
print(merged_inner)

# 2. Left Merge - Keep all from left (soil), add matching
from weather
# Use case: Keep all soil data and check if weather info is
available.
```

```
merged_left = pd.merge(soil_data, weather_data, on='Field_ID',
                                                    how='left')

print("Dataset after Left Merge")
print(merged_left)

# 3. Right Merge - Keep all from right (weather), match with soil
# Use case: Focus on weather sensor plots, and include soil if
present.

merged_right = pd.merge(soil_data, weather_data, on='Field_ID',
                                                    how='right')

print("Dataset after Right Merge")
print(merged_right)

'''
4. Outer Merge - Keep all records, fill missing values with NaN
Use case: Analyze data for all fields (with and without soil and
weather records).
'''

merged_outer = pd.merge(soil_data, weather_data, on='Field_ID',
                                                    how='outer')

print("Dataset after Outer Merge")
print(merged_outer)

print("=====================================")
print("Using of join() method")
print("=====================================")

'''
The join() method combines DataFrames based on the index by
default; so, here is a need to set 'Field_ID' as the index.
It can be Inner Join, Left Join, Right Join, and Outer Join.
'''

# Setting Field_ID as index before joining DataFrames
soil_data.set_index("Field_ID")
weather_data.set_index("Field_ID")

'''
1. Inner Join - Only matching Field_ID entries
Use case: Analyze data only for fields with both soil and weather
records.
'''
inner_join = soil_data.join(weather_data, how='inner',
                            lsuffix='_soil', rsuffix='_weather')

print("Dataset after Inner Join")
print(inner_join)
```

```
# 2. Left Join - Keep all from left (soil), add matching
from weather
# Use case: Keep all soil data and check if weather info is
available.
left_join = soil_data.join(weather_data, how='left',
                           lsuffix='_soil', rsuffix='_weather')
print("Dataset after Left Join")
print(left_join)

# 3. Right Join - Keep all from right (weather), match with soil
# Use case: Focus on weather sensor plots, and include soil if
present.
right_join = soil_data.join(weather_data, how='right',
                            lsuffix='_soil', rsuffix='_weather')
print("Dataset after Right Join")
print(right_join)

'''
4. Outer Join - Keep all records, fill missing values with NaN
Use case: Analyze data for all fields (with and without soil and
weather records).
'''
outer_join = soil_data.join(weather_data, how='outer',
                            lsuffix='_soil', rsuffix='_weather')

print("Dataset after Outer Join")
print(outer_join)

print("======================================")
print("Using of concat() method")
print("======================================")

'''
The concat() method combines DataFrames with same columns as the
original dataframes. Use case: combining two DataFrames either
vertically(stacking rows) or horizontally (combining columns)
to create a single dataset for more comprehensive analysis.
'''
concat_data = pd.concat([soil_data, weather_data])
print("Concatenated Dataset")
print(concat_data)
```

Output of Listing 7.21:

The self-explanatory output of Listing 7.21 is shown below.

```
======================================
Use of  merge(), join(), and concat() Methods
======================================
Original soil dataset
 Field_ID Soil_pH Nitrogen
0 F1         5.5        35
1 F2         6.7        45
2 F3         7.2        49
3 F4         6.5        39
======================================
Original weather dataset
 Field_ID Rainfall_mm Temperature
0 F1         110        24
1 F2         120        26
2 F3          95        23
3 F5          88        25
======================================
Using of merge() method
======================================
Dataset after Inner Merge
 Field_ID Soil_pH  Nitrogen  Rainfall_mm         Temperature
0 F1        5.5      35       110         24
1 F2        6.7      45       120         26
2 F3        7.2      49       95          23
Dataset after Left Merge
 Field_ID Soil_pH Nitrogen Rainfall_mm Temperature
0 F1        5.5      35       110.0     24.0
1 F2        6.7      45       120.0     26.0
2 F3        7.2      49       95.0      23.0
3 F4        6.5      39       NaN       NaN
Dataset after Right Merge
 Field_ID Soil_pH Nitrogen Rainfall_mm Temperature
0 F1        5.5      35.0     110       24
1 F2        6.7      45.0     120       26
2 F3        7.2      49.0     95        23
3 F5        NaN      NaN      88        25
Dataset after Outer Merge
 Field_ID Soil_pH Nitrogen Rainfall_mm Temperature
0 F1        5.5      35.0     110.0     24.0
1 F2        6.7      45.0     120.0     26.0
2 F3        7.2      49.0     95.0      23.0
3 F4        6.5      39.0     NaN       NaN
4 F5        NaN      NaN      88.0      25.0
======================================
Using of join() method
======================================
Dataset after Inner Join
 Field_ID_soil Soil_pH Nitrogen Field_ID_weather Rainfall_mm Temperature
0 F1        5.5        35        F1         110        24
1 F2        6.7        45        F2         120        26
2 F3        7.2        49        F3         95         23
```

```
3 F4      6.5       39        F5        88        25
Dataset after Left Join
 Field_ID_soil Soil_pH Nitrogen Field_ID_weather Rainfall_mm Temperature
0 F1      5.5       35        F1        110       24
1 F2      6.7       45        F2        120       26
2 F3      7.2       49        F3        95        23
3 F4      6.5       39        F5        88        25
Dataset after Right Join
 Field_ID_soil Soil_pH Nitrogen Field_ID_weather Rainfall_mm Temperature
0 F1      5.5       35        F1        110       24
1 F2      6.7       45        F2        120       26
2 F3      7.2       49        F3        95        23
3 F4      6.5       39        F5        88        25
Dataset after Outer Join
 Field_ID_soil Soil_pH Nitrogen Field_ID_weather Rainfall_mm Temperature
0 F1      5.5       35        F1        110       24
1 F2      6.7       45        F2        120       26
2 F3      7.2       49        F3        95        23
3 F4      6.5       39        F5        88        25
========================================
Using of concat() method
========================================
Concatenated Dataset
 Field_ID Soil_pH Nitrogen Rainfall_mm Temperature
0 F1      5.5       35.0      NaN       NaN
1 F2      6.7       45.0      NaN       NaN
2 F3      7.2       49.0      NaN       NaN
3 F4      6.5       39.0      NaN       NaN
0 F1      NaN       NaN       110.0     24.0
1 F2      NaN       NaN       120.0     26.0
2 F3      NaN       NaN       95.0      23.0
3 F5      NaN       NaN       88.0      25.0
```

7.4.2.7 Aggregation and Grouping of Pandas DataFrames

Aggregation and Grouping using Pandas DataFrames is helpful to analyze agricultural data effectively, i.e., for summarizing crop production and calculating statistics such as mean, median, and total production as described in Listing 7.22.

Listing 7.22	Example: *Aggregation* and *Grouping* using Pandas DataFrames

```
# Aggregation and Grouping using Pandas DataFrames
print("=====================================================")
print("# Aggregation and Grouping using Pandas DataFrames")
print("=====================================================")

import pandas as pd

agri_data = {
    'Region': ['North','South','East','North','South','East'],
    'Crop': ['Wheat', 'Rice', 'Corn', 'Barley', 'Wheat', 'Rice'],
    'Production (tons)': [120, 150, 100, 90, 130, 170]
}

agri_df = pd.DataFrame(agri_data)
print("Original soil dataset")
print(agri_df)
print("=====================================")

# Use of groupby() method of DataFrames
# Summarizing total crop production in different regions of a farm
total_regional_production = agri_df.groupby('Region')['Production
                                                        (tons)'].sum()
print("Dataset showing total regional production")
print(total_regional_production)
print("=====================================")

# Summarize avg. production in different farm regions
average_regional_production = agri_df.groupby('Crop')['Production
                                                        (tons)'].mean()
print("Dataset showing average regional production")
print(average_regional_production)
print("=====================================")

# Count of Entries (Group by Region and Crop)
reg_crop_count = agri_df.groupby(['Region', 'Crop']).count()
print("Dataset showing regional crop count")
print(reg_crop_count)
print("=====================================")
```

Output of Listing 7.22:

The self-explanatory output of Listing 7.22 is shown below.

```
=====================================
# Aggregation and Grouping using Pandas DataFrames
=====================================
Original soil dataset
Region crop production (tons)
0 North Wheat        120
1 South Rice         150
2 East Corn          100
3 North Barley        90
4 South Wheat        130
5 East Rice          170
=====================================
Dataset showing total regional production
Region
East                 270
North                210
South                280
Name: Production (tons), dtype: int64
=====================================
Dataset showing average regional production
Crop
Barley                90.0
Corn                 100.0
Rice                 160.0
Wheat                125.0
Name: Production (tons), dtype: float64
=====================================
Dataset showing regional crop count
Production (tons)
Region Crop
East Corn 1
Rice                   1
North Barley           1
Wheat                  1
South Rice             1
Wheat                  1
=====================================
```

7.5 Exercises

Problem 7.1 To ensure agricultural product quality and meet regulatory standards, accurate analysis of nutritional information is essential. A food technologist wants to analyze and explore packaged food data stored in a CSV file named *packaged_food_data.csv* in the following ways:

- Check the overall structure and contents of the packaged food data items
- Handling missing values

- Removing duplicate data entries
- Retrieve all Dairy items with less than 100 Calories
- Summarizing the sum of calories belonging to a specific item category
- Saving DataFrame to CSV file

packaged_food_data.csv

```
ProductID,ProductName,Calories,Sugar(g),Fat(g),Category
1,Granola Bar,120,8,4,Snack
2,Protein Shake,150,12,2,Drink
3,Yogurt,90,6,,Dairy
4,Popcorn,106,1,4,Snack
5,Fruit Juice,,20,0,Drink
6,Yogurt,90,6,,Dairy
```

Problem 7.2 A farm manager wants to manage and analyze agricultural farm machinery data. Considering the data available in the given CSV file (named farm_machinery_data.csv), write a Python program to assist him in doing the following operations:

- Check the overall structure of saved data of agricultural farm machinery
- Handling missing values
- Removing duplicate data entries
- Retrieve specific records, e.g., viewing all tractors or machines made after 2020, or viewing all machines with usage hours more than 2000, or all machines with power greater than 100.
- Save the cleaned and processed DataFrame back into a CSV file for future use.

farm_machinery_data.csv

```
MachineID,MachineType,Brand,Model,Year,Power(HP),FuelType,
UsageHours
M01,Tractor,John Deere,5050E,2019,50,Diesel,1200
M02,Combine Harvester,Claas,Lexion 750,2018,350,Diesel,3000
M03,Plough,Kuhn,Multi-Master,2020,120,Diesel,800
M04,Tractor,Mahindra,Yuvo 575,2021,45,,950
M05,Seeder,John Deere,1590,2017,95,Diesel,1800
M04,Tractor,Mahindra,Yuvo 575,2021,45,,950
M02,Combine Harvester,Claas,Lexion 750,2018,350,Diesel,3000
```

Problem 7.3 An entomologist wants to analyze the available insect data in a CSV file (named *insects_data.csv*). Write a Python program to assist him in doing the following operations:

- Reviewing the dataset to understand its structure
- Retrieving specific records, e.g., viewing only insect names and their lengths, filtering insects by habitat type or region, insects without wings, or insects with lengths longer than 10 mm
- Handling missing data and removing duplicates

insects_data.csv

```
ObservationID,InsectName,Species,Habitat,Length(mm),Wings,Region
ObservationID,InsectName,Habitat,Length(mm),Wings,Region
01,Ladybug,Garden,7,Yes,Europe
02,Monarch Butterfly,Field,45,Yes,North America
03,Honeybee,Apiary,,Yes,Europe
04,Ant,Forest,5,No,Europe
05,Termite,Soil,6,No,North America
03,Honeybee,Apiary,,Yes,Europe
06,Mosquito,Swamp,4,,Asia
```

Chapter 8
Data Science and Python Libraries

8.1 Introduction to Python Libraries

Although the terms *Library* and *Package* are often used interchangeably, there is a subtle distinction: a Python *Package* is typically a collection of modules, while a Python *Library* can be thought of as a collection of related modules and packages. Due of this subtle difference, both have been used interchangeably in the literature. For example, *NumPy* and *Pandas*, which we previously referred to as packages, are also referred to as libraries. Thus, whether we call them packages or libraries, the important thing is to focus on how to use them, especially in the context of agriculture, rather than getting caught up in the terminology.

Considering the context of data science implications in agriculture, there are many Python libraries available today. *Matplotlib* and *Scikit*-learn are two popular Python libraries discussed in this chapter. Occasionally, when working with data, you need to visualize trends in datasets and apply machine learning (ML) techniques to make predictions or uncover patterns. For such cases, *Matplotlib* and *Scikit-learn* can be used. *Matplotlib* is a widely used library for creating data visualizations, including simple 2D graphs and more complex, interactive, or animated charts (making it easier to understand data visually). On the other hand, *Scikit-learn* is also a free and open-source library that provides simple tools for building and using machine learning models, which is used for tasks such as classification, prediction, or clustering. It is important to mention here that these libraries are designed to work with Python Compound data structures (i.e., List, Tuple, Dictionary, and Set), NumPy arrays, and Pandas DataFrames.

M. A. Iqbal, *Python for Agriculturists*,
https://doi.org/10.1007/978-3-032-01442-9_8

8.2 Python Matplotlib Library

Matplotlib (developed by John D. Hunter) is an open-source graph-plotting Python library that is used mainly for creating static, interactive, and animated visualizations. It is widely employed in data science and scientific computing for generating high-quality plots and charts (line graphs, bar graphs, pie graphs, scatter graphs, etc.) from available datasets. In the context of agriculture, Matplotlib becomes an essential tool for visually analyzing trends and patterns in agricultural data. Farmers, researchers, and agri-tech professionals can use it to visualize crop yield over time, rainfall and temperature patterns in agricultural datasets, pest infestations in particular regions, soil nutrients proportion at different farms, and so on. These visual insights help agriculturists in identifying problems early, communicating findings effectively, and making data-driven decisions in digital agriculture research.

8.2.1 Plot Line Graphs with Matplotlib Using Python Data Structures and DataFrames

Based on agricultural scenarios, graph plotting using Matplotlib with data stored in Python data structures (i.e., List, Tuple, Dictionary, and Set) is demonstrated from Listings 8.1–8.5. The code in these listings includes comments explaining various graph customization features (i.e., titles, labels, legends, gridlines, colors, markers, and line styles) to enhance the reader's understanding and make the concepts easier to follow.

8.2.1.1 Line Graph Using Python List Data

Listing 8.1	Example creating graphs using matplotlib with data stored in *list* data structure of Python

```
'''
=============================================================
Step 1: Importing Pyplot module of Matplotlib library is
essential to draw graphs and charts because it provides methods
for creating visualizations.
=============================================================
'''
import matplotlib.pyplot as plt
'''
```

```
======================================================
Step 2: Consider crop yield data over years stored in
Python Lists
======================================================
'''
years = [2020, 2021, 2022, 2023, 2024]
wheat_yield = [2.5, 2.7, 3.0, 3.2, 3.5]
rice_yield = [2.0, 2.4, 2.8, 3.1, 3.3]
'''
======================================================
Step 3: Use of plot method() of pyplot module for plotting
line graphs
The plot() method of matplotlib.pyplot module is used to create
line graphs. It takes several arguments that help control the
appearance and behavior of the plot. Below is the brief
description of the most commonly used arguments (also used in
this example).

- 1st and 2nd arguments represents the data points on the x-axis
and y-axis, respectively. (In this example, years represents
x-axis, and (crop) yield represents y-axis)
- label: Assigns a name to the line, useful when creating
legends.
- color: Sets the color of the line.
- linestyle or ls: Defines the style of the line.
- linewidth or lw: Controls the thickness of the line.
Common Options: '-' (solid), '--' (dashed), ':' (dotted), '-.'
(dash-dot)
- marker: Defines the style of data point markers.
Common Options: marker='o' (circle), 's' (square), 'H'(hexagon)
- markersize or ms: Sets the size of the markers.
======================================================
'''

# Use plot method() for plotting wheat yield
plt.plot(years, wheat_yield, label='Wheat', color='green',
ls='--', lw = 3, marker='o')

# Use plot method() for plotting rice yield
plt.plot(years, rice_yield, label='Rice', color='brown', ls='-.',
lw = 3, marker='s')
'''
```

```
=====================================================
Step 4: Use various pyplot methods (i.e., title(), xlabel(),
ylabel(), etc.) to customize the plotted graph.
=====================================================
'''

# Use title() method for setting the graph title
plt.title('Crop Yield Trends (2020-2024)')

# Use of xlabel() method to label x-axis
plt.xlabel('Year')

# Use of ylabel() method to label y-axis
plt.ylabel('Yield (tons/hectare)')

# Use legend() method for setting graph legend
plt.legend()

# Use grid() method for setting grid lines
plt.grid(True)
'''
=====================================================
Step 5 (optional): Use xticks() method of of pyplot module with
the range function to force x-axis ticks to be displayed as
integer values. Without using the xticks() method in this way,
x-axis ticks will be displayed as float values.
=====================================================
'''
plt.xticks(range(min(years), max(years)+1))
# Use show() method to make the graph visible
plt.show()
```

Output of Listing 8.1

The output of Listing 8.1 is shown in Fig. 8.1.

The graph plotting process has been clearly explained through inline comments in Listing 8.1. It is important to carefully follow each step described in the comments, as they provide a detailed, step-by-step explanation. This approach helps readers grasp the concepts more effectively and enhances their overall understanding.

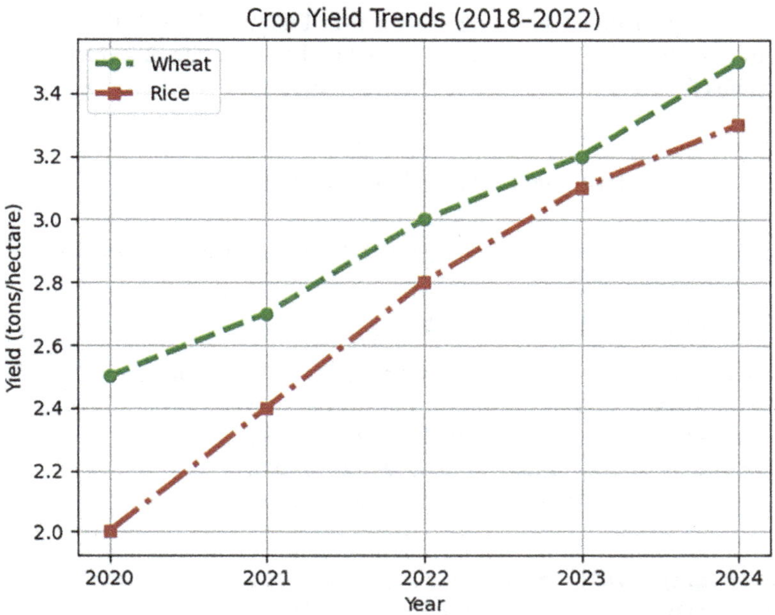

Fig. 8.1 Graph of (wheat and rice) crop yield against years

8.2.1.2 Line Graph Using Python Tuple Data

Listing 8.2	Example creating graphs using matplotlib with data stored in *Tuple* Data Structure of Python

```
import matplotlib.pyplot as plt

# Crop yield over years stored in Python Tuple

years = (2020, 2021, 2022, 2023, 2024)
wheat_yield = (2.5, 2.7, 3.0, 3.2, 3.5)
rice_yield = (2.0, 2.4, 2.8, 3.1, 3.3)

plt.plot(years, wheat_yield, label='Wheat', color='green',
ls='--', lw = 3, marker='o')
plt.plot(years, rice_yield, label='Rice', color='brown', ls='-.',
lw = 3, marker='s')

plt.title('Crop Yield Trends (2020-2024)')
plt.xlabel('Year')
plt.ylabel('Yield (tons/hectare)')
plt.legend()
plt.grid(True)
plt.xticks(range(min(years), max(years)+1))
plt.show()
```

The code in Listing 8.2 is similar to that in Listing 8.1, with the only change being the use of Python Tuple instead of Python List. Despite this change, the output remains the same as the graph displayed in Fig. 8.1.

8.2.1.3 Line Graph Using Python Dictionary Data

Listing 8.3	Example creating graphs using matplotlib with data stored in *Dictionary* Data Structure of Python

```
import matplotlib.pyplot as plt

# Crop yield over years stored in Python Dictionary
crop_yield_data = {
    'Years': [2020, 2021, 2022, 2023, 2024],
    'Wheat_Yield': [2.5, 2.7, 3.0, 3.2, 3.5],
    'Rice_Yield': [2.0, 2.4, 2.8, 3.1, 3.3]
}

plt.plot(crop_yield_data['Years'], crop_yield_data['Wheat_
Yield'], label='Wheat', color='green', ls='--', lw = 3,
marker='o')
plt.plot(crop_yield_data['Years'], crop_yield_data['Rice_Yield'],
label='Rice', label='Rice', color='brown', ls='-.', lw = 3,
marker='s')

plt.title('Crop Yield Trends (2020-2024)')
plt.xlabel('Year')
plt.ylabel('Yield (tons/hectare)')
plt.legend()
plt.grid(True)
plt.xticks(range(min(years), max(years)+1))
plt.show()
```

The code in Listing 8.3 is similar to that in Listing 8.1, with a few changes, including the use of a Python Dictionary instead of a Python List and related changes for *x*-axis and *y*-axis arguments of the plot method. Despite these changes, the output remains the same as the graph displayed in Fig. 8.1.

8.2.1.4 Line Graph Using Python Set Data

Listing 8.4	Example creating graphs using matplotlib with data stored in *Set* Data Structure of Python

```
import matplotlib.pyplot as plt

# Data stored in Python sets
years_set = {2020, 2021, 2022, 2023, 2024}
wheat_yield_set = {2.5, 2.7, 3.0, 3.2, 3.5}
rice_yield_set = {2.0, 2.4, 2.8, 3.1, 3.3}

# Convert sets to sorted lists for plotting
years = sorted(list(years_set))
wheat_yield = sorted(list(wheat_yield_set))
rice_yield = sorted(list(rice_yield_set))

plt.plot(years, wheat_yield, label='Wheat', color='green',
ls='--', lw = 3, marker='o')
plt.plot(years, rice_yield, label='Rice', color='brown', ls='-.',
lw = 3, marker='s')
plt.title('Crop Yield Trends (2020-2024)')
plt.xlabel('Year')
plt.ylabel('Yield (tons/hectare)')
plt.legend()
plt.grid(True)
plt.xticks(range(min(years), max(years)+1))
plt.show()
```

The code in Listing 8.4 is similar to that in Listing 8.1, with a few changes, including the use of Python Sets instead of Python Lists and converting these sets into sorted lists before plotting the graph. Despite these changes, the output remains the same as the graph displayed in Fig. 8.1.

8.2.1.5 Line Graph Using Python Pandas DataFrames

To plot Graphs with Matplotlib using Python DataFrames, consider that the data used in the above examples is stored in a CSV file named *crop_yield_data.csv* (as shown below). Listing 8.5 demonstrates the use of graph creation using Python Pandas DataFrame.

crop_yield_data.csv

```
Years,Wheat_Yield,Rice_Yield
2020,2.5,2
2021,2.7,2.4
2022,3,2.8
2023,3.2,3.1
2024,3.5,3.3
```

Listing 8.5	Example creating graphs using matplotlib with DataFrames (for data stored in CSV file)

```
# Crop yield over years stored in CSV file
import pandas as pd
import matplotlib.pyplot as plt

# Read crop yield data from CSV file
crop_yield_data = pd.read_csv('crop_yield_data.csv')

plt.plot(crop_yield_data['Years'], crop_yield_data['Wheat_
Yield'], label='Wheat', color='green', ls='--', lw = 3,
marker='o')

plt.plot(crop_yield_data['Years'], crop_yield_data['Rice_Yield'],
label='Rice', color='brown', ls='-.', lw = 3, marker='s')
plt.title('Crop Yield Trends (2020-2024)')
plt.xlabel('Year')
plt.ylabel('Yield (tons/hectare)')
plt.legend()
plt.grid(True)
plt.
xticks(range(min(crop_yield_data['Years']), max(crop_yield_
data['Years'])+1))
plt.show()
```

The code in Listing 8.5 is similar to that in Listing 8.1, with a few changes, including the use of Python Pandas (instead of using any Python data structures) for reading of CSV files to store data in Python Pandas, and DataFrame, and related changes for x-axis and y-axis arguments of the plot method. Despite these changes, the output remains the same as the graph displayed in Fig. 8.1.

8.2.1.6 Twin Line Graph

Sometimes, creating a twin graph in Matplotlib is quite helpful, especially when comparing two different datasets that share the same x-axis but require separate y-axis. For instance, comparing crop yield and rainfall over the same set of years is effectively represented using a twin line graph, as demonstrated in Listing 8.6.

Listing 8.6	Example demonstrating the creation of a Twin graph using matplotlib

```
import matplotlib.pyplot as plt

# Six months and related weather data stored in Lists
months = ['Jan', 'Feb', 'Mar', 'Apr', 'May', 'Jun']
temperature = [13, 19, 24, 25, 28, 31] # in °C
rainfall = [70, 55, 43, 33, 25, 15] # in mm
'''
To create a Twin graph, use the subplots() method without
arguments that returns a figure and a single axis.
'''
fig, left_axis = plt.subplots()

# Use set_xlabel() method to set the label of the x-axis.
left_axis.set_xlabel('Month')

# Use set_ylabel() to set label & color of left y-axis left_axis.
set_ylabel('Temperature (°C)', color='red')

# Use plot() to plot temperature readings over months.
left_axis.plot(months, temperature, color='red', marker='X',
label='Temperature')
'''
To create a second y-axis (right_axis) that shares the same
x-axis, use twinx() method to show two different scales on
one plot.
'''
right_axis = left_axis.twinx()

# Use set_ylabel() to set label & color of left y-axis
right_axis.set_ylabel('Rainfall (mm)', color='blue')

# Use plot() to plot rainfall readings over months.
right_axis.plot(months, rainfall, color='blue', marker='H',
label='Rainfall')

plt.title('Temperature and Rainfall Patterns Over Time')
plt.grid(True)
plt.show()
```

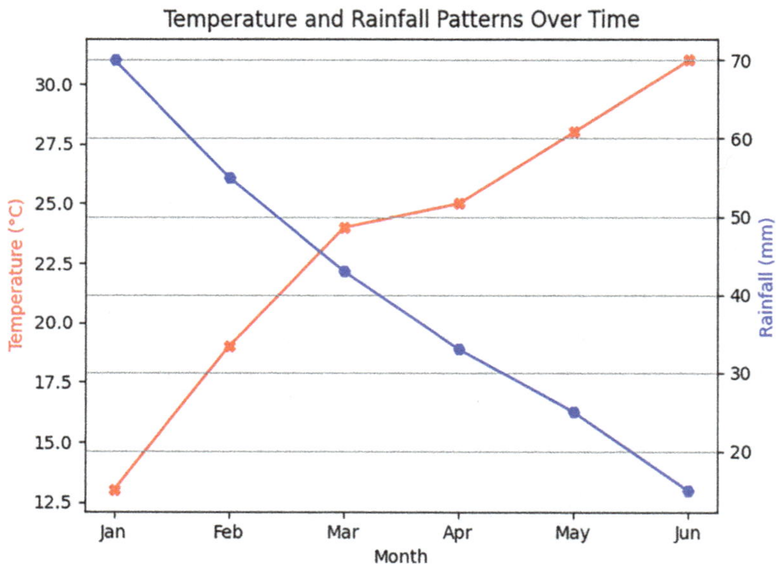

Fig. 8.2 Twin graphs showing temperature and rainfall patterns over time

Output of Listing 8.6
The output of Listing 8.6 is shown in Fig. 8.2.

8.2.2 Plotting Bar Graphs with Matplotlib

Matplotlib also supports the creation of Bar graphs as explained in Listings 8.7–8.11.

8.2.2.1 Plotting Vertical Bar Graph

Listing 8.7	Example demonstrating the creation of a bar graph using matplotlib

```
import matplotlib.pyplot as plt
import pandas as pd

# Read milk yield data from CSV file
milk_yield_df = pd.read_csv('milk_yield_data.csv')
```

```
'''
Use the bar() method of the pyplot module for plotting bar
graphs. It takes several arguments that help control the
appearance and behavior of the graph.
Below is a brief description of the most commonly used arguments
(also used in this example).

- 1st argument represents the data points on the x-axis
- 2nd argument represents the height of each bar
- (optional) width argument to adjust width (default =0.8)
- (optional) color to fill color in bars
- (Optional) align bars to 'center' (default) or 'edge' of
x-ticks.
- (optional) edgecolor to color the edges (borders) of the bars.
'''

plt.bar(milk_yield_df['CowID'], milk_yield_df['MilkYield(L)'],
width = 0.5, color=['skyblue', 'lightgreen', 'coral', 'plum'],
edgecolor = 'black')

# Customizations to the plotted graph.
plt.title('Cow Milk Production by Breed')
plt.xlabel('Breed')
plt.ylabel('Milk Production (Litres)')
plt.grid(axis='y')
plt.show()
```

Output of Listing 8.7

The output of Listing 8.7 is shown in Fig. 8.3.

8.2.2.2 Plotting Horizontal Bar Graph

Listing 8.8	Example demonstrating the creation of horizontal bar graph using matplotlib

```
import matplotlib.pyplot as plt
import pandas as pd

# Read crop milk yield data from CSV file
```

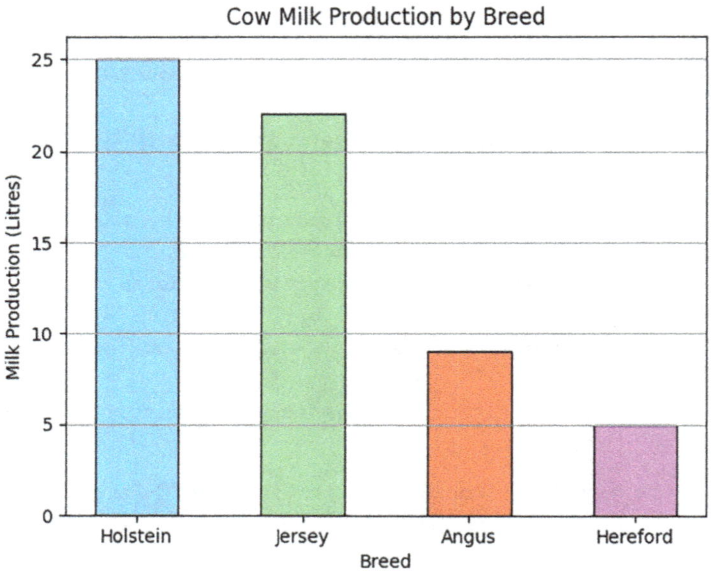

Fig. 8.3 Milk production of different cow breeds

```
milk_yield_df = pd.read_csv('milk_yield_data.csv')
'''
Use bar() method of pyplot module for plotting horizontal
bar graphs
It takes several arguments that help control the appearance and
behavior of the graph.
Below is the brief description of the most commonly used
arguments (also used in this example).

- 1st argument represents the data points on the y-axis
- 2nd argument represents the width of each bar
- (optional) height argument to adjust width (default =0.8)
- (optional) color to fill color in bars
- (Optional) align bars to 'center' (default) or 'edge' of
x-ticks.
- (optional) edgecolor to color the edges (borders) of the bars.
'''
plt.barh(milk_yield_df['CowID'], milk_yield_df['MilkYield(L)'],
height = 0.5, color=['skyblue', 'lightgreen', 'coral', 'plum'],
edgecolor = 'black')

# Customizations to the plotted graph.
plt.title('Cow Milk Production by Breed')
```

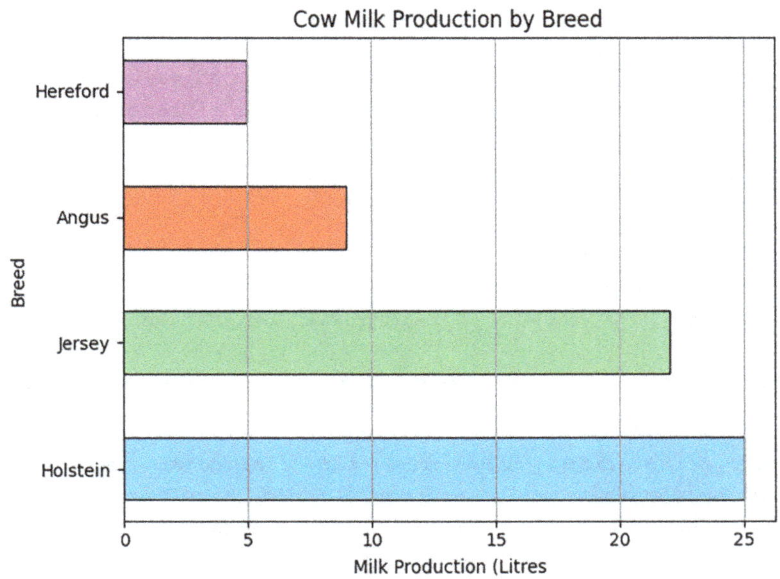

Fig. 8.4 Milk production of different cow breeds in horizontal bar graph

```
plt.xlabel('Milk Production (Litres')
plt.ylabel('Breed')
plt.grid(axis='x')
plt.tight_layout()
plt.show()
```

Output of Listing 8.8

The output of Listing 8.8 is shown in Fig. 8.4.

8.2.2.3 Plotting Stacked Bar Graph

Listing 8.9	Example demonstrating the creation of a stacked graph using matplotlib

```
import matplotlib.pyplot as plt

crops = ['Wheat', 'Rice', 'Maize']

'''
Three lists (below) represent nutrient values (correspond
positionally) for each crop.
For example, Wheat has: Nitrogen: 30, Phosphorus: 20,
Potassium: 15.
'''
nitrogen = [30, 40, 35]
phosphorus = [20, 25, 20]
```

```
potassium = [15, 10, 20]
'''
Creation of the first layer of bars for nitrogen values where
each crop gets a bar whose height equals the nitrogen amount.
'''
plt.bar(crops, nitrogen, label='Nitrogen')

'''
Creation of a stacked bar for phosphorus on top of nitrogen.
The bottom=nitrogen means that the phosphorus bars start where
the nitrogen bars end and therefore will be stacked visually.
'''
plt.bar(crops, phosphorus, bottom=nitrogen, label='Phosphorus')

'''
Further, for stacking potassium on top of nitrogen + phosphorus,
correct bottom (base)value is required. So, here zip function of
Python is used that after taking iterable as arguments returns a
zip object. The returned zip object is an iterator of tuples
where the corresponding data elements in each passed iterator are
paired together. Next, list comprehension [n + p for n, p in
zip(...)] has been used to calculate correct bottom base for
potassium by adding each nitrogen + phosphorus value.
For example:
1 - zip(nitrogen, phosphorus) pairs up corresponding values from
both lists:
[(30, 20), (40, 25), (35, 20)]

2 - [30 + 20, 40 + 25, 35 + 20] → bottom_vals = [50, 65, 55]
bottom_vals list now has the correct base (bottom) for stacking
the potassium bars on top of nitrogen + phosphorus.
'''
bottom_vals = [n + p for n, p in zip(nitrogen, phosphorus)]

'''
Creation of a stacked bar for potassium on top of nitrogen +
phosphorus.
The bottom=bottom_vals means that the potassium bars start where
the nitrogen+phosphorus bars end and therefore will be stacked
visually.
'''
plt.bar(crops, potassium, bottom=bottom_vals, label='Potassium')

# Customizations to the plotted graph.
plt.xlabel('Crops')
plt.ylabel('Fertilizer (kg/ha)')
plt.title('Fertilizer Usage by Crop')
plt.legend(loc='upper left')
plt.show()
```

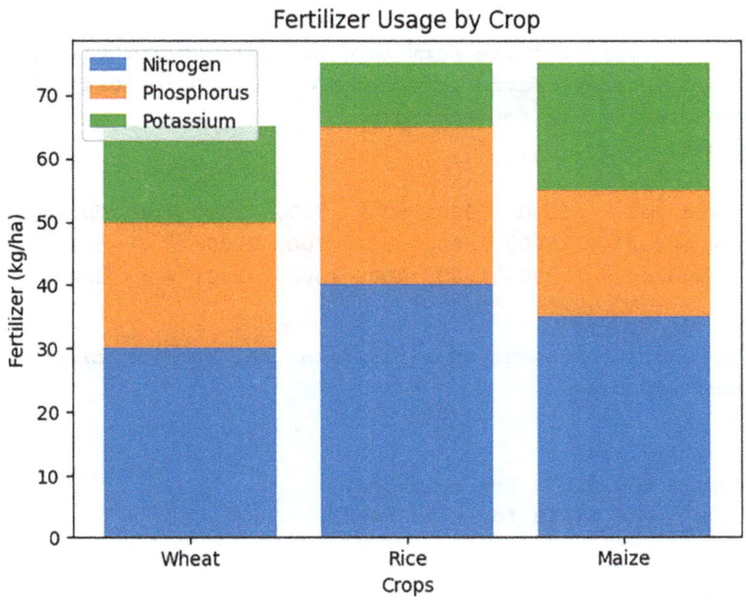

Fig. 8.5 Stacked graph representing fertilizer consumption by different crops

Output of Listing 8.9

The output of Listing 8.9 is shown in Fig. 8.5.

8.2.2.4 Plotting Multiple Bars in a Bar Graph

Listing 8.10	Example demonstrating the creation of multiple bars in a single graph using matplotlib

```
import matplotlib.pyplot as plt

'''
Set the width of each bar in the plot (as bars for different farms
expenses will appear side by side for each year, a smaller width
helps prevent overlap.

As bar width required same for all bars, so declared it as a
constant
'''
BAR_WIDTH = 0.25
```

```
'''
Lists below represent the farm expenses data (for three farms
i.e., farm1, farm2, farm3 across 5 years (2020 to 2024). Each
list contains values corresponding
to the same years.
'''
farm1_expenses = [1500, 1000, 500, 1300, 2100] # in dollars
farm2_expenses = [1200, 1300, 300, 1300, 2300] # in dollars
farm3_expenses = [1300, 1100, 400, 1300, 3100] # in dollars

'''
Creating lists representing x-positions for bars of farm1, farm2,
and farm3_expenses.
'''
bar1 = [0,1,2,3,4]
bar2 = [x + BAR_WIDTH for x in bar1]
bar3 = [x + BAR_WIDTH for x in bar2]

# Plotting of bar graphs using bar() method of pyplot
plt.bar(bar1, farm1_expenses, color ='r', width = BAR_WIDTH,
        edgecolor ='grey', label ='Farm 1')
plt.bar(bar2, farm2_expenses, color ='g', width = BAR_WIDTH,
        edgecolor ='grey', label ='Farm 2')
plt.bar(bar3, farm3_expenses, color ='b', width = BAR_WIDTH,
        edgecolor ='grey', label ='Farm 3')

# Customizations to the plotted graph

# Setting axes labels' font size/weight
plt.xlabel('Years', fontweight ='bold', fontsize = 15)
plt.ylabel('Total Farm Expenses ($)', fontweight ='bold',
fontsize = 15)

# Setting x-axis labels under middle bar of each group
plt.xticks([0.25, 1.25, 2.25, 3.25, 4.25],['2020', '2021',
'2022', '2023', '2024'])
plt.grid(axis='y')
plt.legend()
plt.show()
```

Output of Listing 8.10

The output of Listing 8.10 is shown in Fig. 8.6.

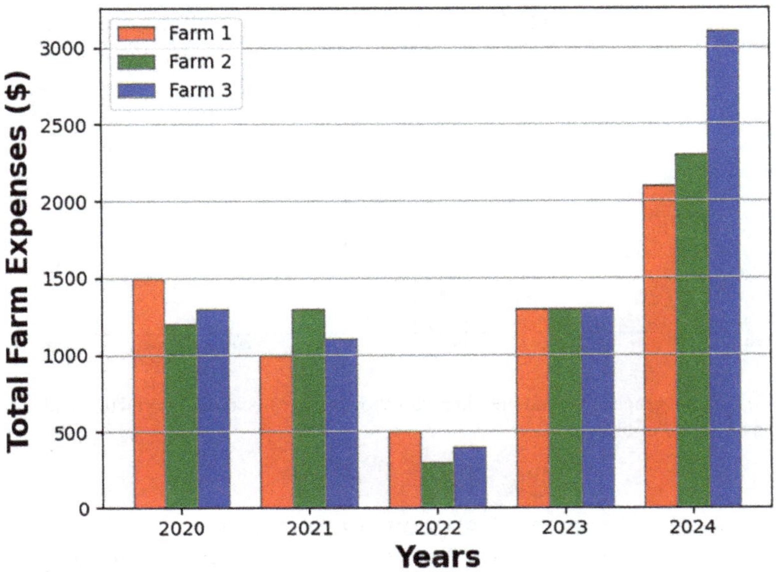

Fig. 8.6 Multiple bars showing farms' expenses over different years

8.2.3 Plotting Multiple Graphs in Single Plot

Listing 8.11	Example demonstrating the creation of multiple graph in a single plot using matplotlib

```
import matplotlib.pyplot as plt
import pandas as pd

# Read milk and crop yield data from CSV files
milk_yield_df = pd.read_csv('milk_yield_data.csv')
crop_yield_df = pd.read_csv('crop_yield_data.csv')
# Figure with 1 row, 2 cols., 12"wide 5" tall
fig, axis = plt.subplots(1, 2, figsize=(12, 5))

# Subplot 1: Bar graph - milk yield by breed
axis[0].bar(milk_yield_df['CowID'], milk_yield_
df['MilkYield(L)'], width = 0.5, color=['skyblue', 'lightgreen',
        'coral', 'plum'], edgecolor = 'black')
axis[0].set_title('Cow Milk Production by Breed')
axis[0].set_xlabel('Breed')
axis[0].set_ylabel('Milk Production (Litres)')
axis[0].grid(axis='y')
```

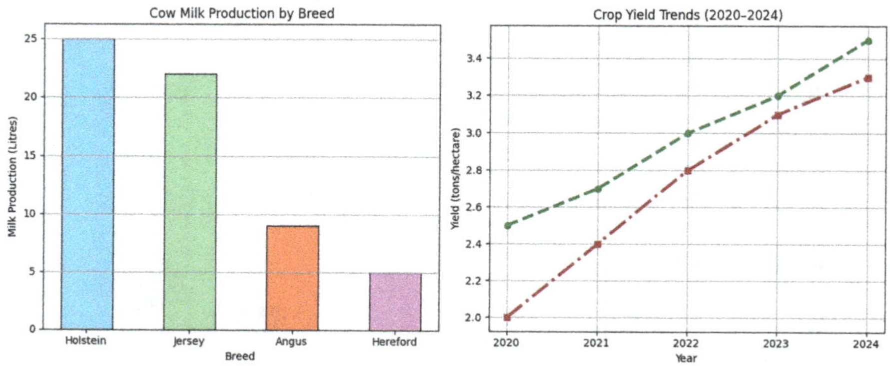

Fig. 8.7 Multiple graphs (bar and line) in p Plot showing livestock milk production and crop yield trends in an agricultural farm

```
# Subplot 2: Line graph - crop yield over years
axis[1].plot(crop_yield_df['Years'], crop_yield_df['Wheat_
Yield'], label='Wheat', color='green', ls='--', lw = 3,
marker='o')
axis[1].plot(crop_yield_df['Years'], crop_yield_df['Rice_Yield'],
label='Rice', color='brown', ls='-.', lw = 3, marker='s')
axis[1].set_title('Crop Yield Trends (2020-2024)')
axis[1].set_xlabel('Year')
axis[1].set_ylabel('Yield (tons/hectare)')
axis[1].grid(True)
plt.
xticks(range(min(crop_yield_df['Years']), max(crop_yield_
df['Years'])+1))
plt.tight_layout()
plt.show()
```

Output of Listing 8.11

The output of Listing 8.11 is shown in Fig. 8.7.

8.2.4 Plotting Pie Graphs

Listing 8.12	Example demonstrating the creation of pie graphs using matplotlib

```
import matplotlib.pyplot as plt

# Crop names and related production data in lists
crop_types = ['Wheat', 'Rice', 'Maize', 'Oat']
crop_production = [40, 25, 20, 15] # tonnes/acres
```

```
# Figure with 1 row, 2 cols., 12" wide and 5" tall
fig, axis = plt.subplots(1, 2, figsize=(12, 5))

# Creation of a pie chart
'''
The pie() method of matplotlib.pyplot module is used to create
pie graphs.It takes several arguments that help control the
appearance and behavior of the plot. Below is the brief
description of the most commonly used arguments.
- 1st argument must be a sequence of values that represent the
size of each wedge.
- 2nd (optional) argument should list of labels for each wedge
- 3rd (optional) argument should be autopct representing a string
or function to label edges with
numeric values e.g., "%.1f%%" to show percentage.
- 4th (optional) argument should be startangle in degrees showing
the start of pie chart.
(Commonly set to 90 to start at the top)
- 5th (optional) argument is shadow to add a shadow for
visual effects
- 6th (optional) argument is explode that is the list of values
to pull out wedges from the pie for
emphasis. For example, explode = [0, 0.1, 0, 0] means that wedge
representing the second value in
crop production list will be pulled out from pie chart.
- 7th (optional) argument is list of colors to apply to each
wedge of pie chart. If not provided,
Matplotlib uses its default color cycle
'''

# Subplot 1: Pie graph
axis[0].pie(crop_production, labels=crop_types, autopct='%.0f%%',
startangle=90)
axis[0].set_title('Crop Production on the Farm')
axis[1].pie(crop_production, labels=crop_types,
autopct='%0.2f%%', startangle=60, explode=[0, 0.2, 0, 0],
shadow=True)
axis[1].set_title('Crop Production on the Farm')
plt.show()
```

Output of Listing 8.12

The output of Listing 8.12 is shown in Fig. 8.8.

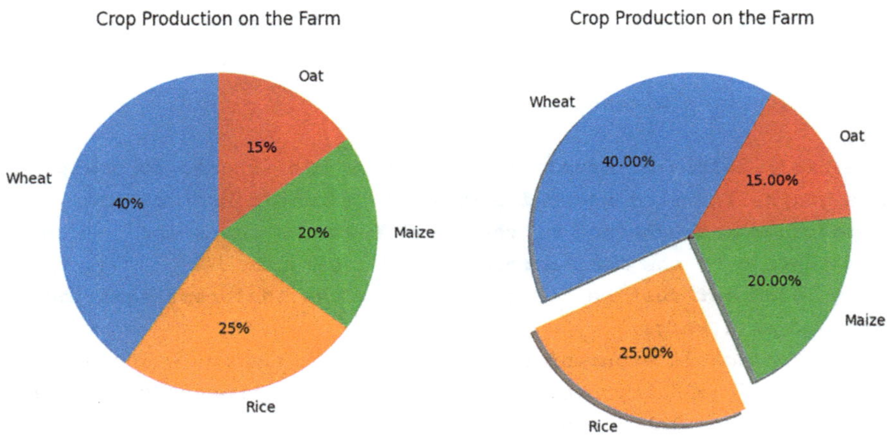

Fig. 8.8 Two versions of pie graphs of the same data showing crop production on the farm

8.2.5 Plotting Scatter Graph

Listing 8.13	Example demonstrating the creation of scatter graphs using matplotlib

```
import matplotlib.pyplot as plt

# Irrigation and corresponding recorded plant growth data
irrigation_levels = [100, 200, 300, 400, 500, 600] # L/week
plant_growth = [10, 15, 18, 22, 20, 17] # cm

'''
Creation of scatter plot to graphically represent the impact of
irrigation level on plant growth
'''
plt.scatter(irrigation_levels, plant_growth, color='green',
marker='o')

# Customizations to the plotted graph
plt.title('Irrigation vs Plant Growth')
plt.xlabel('Irrigation Level (liters/week)')
plt.ylabel('Plant Growth (cm)')
plt.grid(True)
plt.show()
```

Output of Listing 8.13

The output of Listing 8.13 is shown in Fig. 8.9.

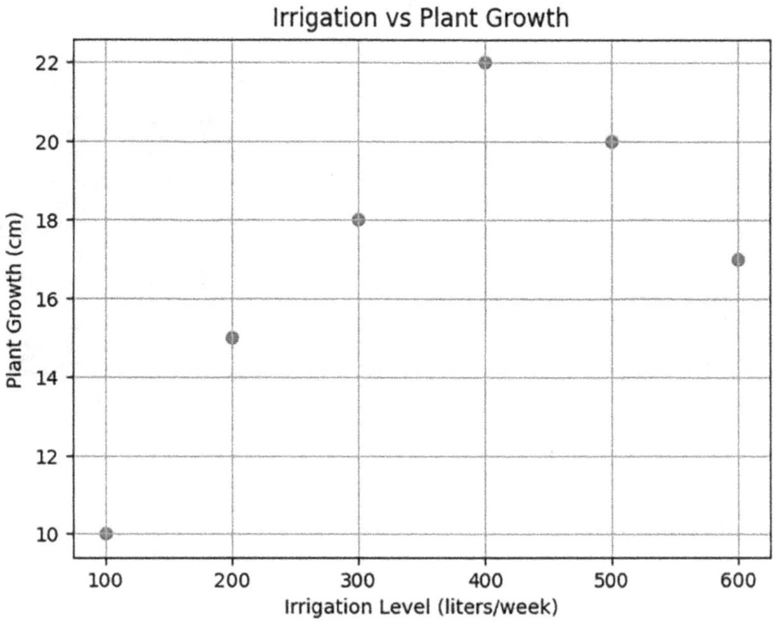

Fig. 8.9 Scattered graph showing relationship between irrigation level and plant growth on an agriculture farm

8.2.6 Plotting Histogram Graph

Listing 8.14	Example demonstrating the creation of a histogram graph using matplotlib

```
import matplotlib.pyplot as plt

# Sample crop yield data in tons/hectare
crop_yields = [2.5, 3.0, 3.1, 3.3, 3.5, 3.5, 3.7, 4.0, 4.2, 4.5,
4.5, 4.7, 5.0, 5.1, 5.5]

# Figure with 1 row, 2 cols., 12" wide & 5" tall
fig, axis = plt.subplots(1, 2, figsize = (12, 5))

# Creating histograms using hist() method
axis[0].hist(crop_yields, bins=3, color='green',
edgecolor='black')
axis[0].set_title('Crop Yield Histogram using 3 Groups')
axis[0].set_xlabel('Groups')
axis[0].set_ylabel('Frequency')
axis[1].hist(crop_yields, bins=6, color='red', edgecolor='black')
axis[1].set_title('Crop Yield Histogram using 6 Groups')
axis[1].set_xlabel('Group Range')
axis[1].set_ylabel('Frequency')

plt.show()
```

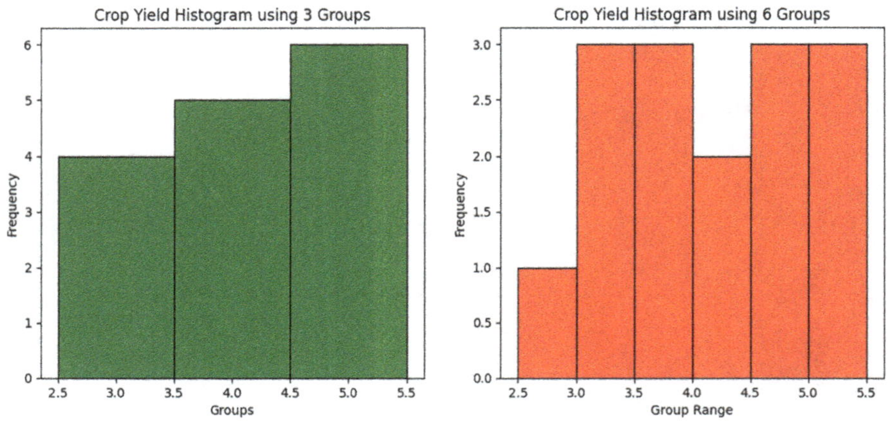

Fig. 8.10 Histograms showing crop yield distribution using 3 and 6 groups

Output of Listing 8.14

The output of Listing 8.14 is shown in Fig. 8.10.

8.3 Scikit-learn Python Library

Scikit-learn is an open-source Python library built on top of NumPy and Matplotlib, mainly used to implement machine learning algorithms. It provides simple and efficient tools for data preprocessing and training machine learning models. In the agriculture context, scikit-learn is extremely useful for tasks such as crop yield prediction, soil classification, disease detection and classification, and smart agriculture applications. For instance, agricultural researchers can use it to train predictive models using historical weather, soil/crop data, and so on, helping farmers make data-driven decisions. Scikit-learn supports a wide range of supervised and unsupervised machine learning algorithms (e.g., linear regression, decision trees, Support Vector Machines (SVMs), K-means), making it ideal for analyzing complex agricultural datasets and building intelligent farming systems.

8.3.1 Key Steps for Scikit-learn Machine Learning Project

A typical Scikit-learn machine learning project in Python involves the following key steps:

1. *Step 1: Setting up the environment*: By importing all the necessary libraries and packages required for data loading, preprocessing, model training, and result visualization.

2. *Step 2: Data preparation*: This includes

    (a) Loading the dataset using Pandas' built-in methods.
    (b) Exploring the dataset using Pandas' and Matplotlib's built-in methods.
    (c) Cleaning the data by handling missing values, detecting and treating outliers, and resolving any inconsistencies.
    (d) Data preprocessing by scaling feature(s) using the StandardScalar method of Scikit-learn's preprocessing module.
    (e) Splitting the dataset into training and testing subsets using the *train_test_split()* method of the Scikit-learn library.

3. *Step 3: Model selection, training, and predictions*: Choose an appropriate machine learning algorithm based on the type of problem, i.e., regression analysis, classification, or clustering. After model selection, use the model to train the selected model using the *fit()* method on the training dataset. Ultimately, the trained model can be used to predict outcomes on the test dataset.
4. *Step 4: Model evaluation*: Assess the model's performance using suitable metrics such as accuracy, precision, recall, F1-score, or confusion matrix.
5. *Step 5: Saving the rained model*: Save the final model for future reuse using Python's *joblib* module.

Considering the applications of Scikit-learn in the agricultural domain, all the key machine learning steps have been demonstrated using the following case study (with its corresponding dataset).

8.3.2 Case Study 8.1: Predicting Crop Yield Using Environmental and Agricultural Inputs

In modern agriculture, the accurate estimation or prediction of crop yield is essential for a farm's resource planning and management. The prediction of crop yield is dependent on several critical environmental factors and agricultural inputs, i.e., rainfall, temperature, fertilizer, and pesticide applications. For this purpose, machine learning techniques can be used to build predictive models not only to understand the impact of known feature(s) on crop yield but also to identify which feature(s) contribute most to yield prediction. Considering the dataset shown in Table 8.1, use machine learning techniques to build predictive models that estimate crop yield using:

- Single feature at a time, for example,
    - Predict crop yield based on rainfall
    - Predict crop yield based on temperature
    - Predict crop yield based on fertilizer usage
- Multiple features collectively, for example,
    - Predict crop yield using a combination of rainfall, temperature, and fertilizer

Table 8.1 Crop yield dataset

Crop_id	Rainfall	Temperature	Fertilizer	Pesticide	Crop_Yield	Yield_status
Crop1	849.67	21.17	127.16	2.59	15.14	Good
Crop1	786.17	23.16	131.22	2.72	16.06	Better
Crop1	864.77	23.31	141.66	3.37	17.34	Optimal
Crop1	952.3	22.4	141.08	3.31	17.48	Optimal
Crop1		23.68	92.45	2.99	15.6	Good
Crop1	776.59	24.81	101.24	3.06	16.69	Better
Crop1	957.92	27.77	130.3	3.64	16.85	Better
Crop1	876.74	24.35	130.28	2.7	17.07	Optimal
Crop1	753.05	24.52	130.3	3.27	16.92	Better
Crop1		23.85		2.9	18.57	Optimal
Crop2	753.66	20.16	131.42	2.89	15.75	G
Crop2	753.43	23.95	142.71	3.55	16.52	B
Crop2	824.2	10	139.08	5.43	16.58	B
Crop2	608.67		133.03	8.03	15.87	G
Crop2	1500	23.62	177	3.65	15.99	G
Crop2	743.77	11	135.18	3.01	16.43	B
Crop2		23.93	104.54		16.71	B
Crop2	831.42	21.66	115.26	2.84	15.94	G
Crop2	709.2		110.29	3.16	16.54	B
Crop2	658.77	25.5	121.64	7.56	16.08	B
Crop3	946.56	25.58	166.29	3.05	19.44	Optimal
Crop3	777.42	22.18		3.3	16.22	Better
Crop3	806.75	26.81	133.73	2.59	16.34	Better
Crop3	1200	21.2	87.75		15.41	Good
Crop3	745.56		110.56	2.5	16.46	Better
Crop3		28.38	141.78	2.39	16.76	Better
Crop3	684.9	22.02	180	3.58	15.26	Good
Crop3	837.57	22.87	98.45	3.4	16.56	Better
Crop3	739.94	24.2	105.69	3.31	16.76	Better
Crop3	770.83	22.99	133.59	3.31	15.7	Good

Solution: Considering the typical key steps involved in a machine learning project using the Scikit-Learn library (along with associated packages and modules) have been demonstrated in Listings 8.15–8.22.

8.3.2.1 Step 1: Setting Up the Environment

Import all the necessary libraries and packages required for data loading, preprocessing, model training, and result visualization.

Listing 8.15	Step 1: Importing the required Python packages and libraries

```
print("==========================================================")
print("Step1: Importing required Python pacakages and libraries")
print("==========================================================")

'''
Importing essential Python packages and libraries for data
loading, preprocessing, model training, and result visualization.
'''

import numpy as np        # For results returned as numpy array
import pandas as pd       # For data loading and preprocessing
import matplotlib.pyplot as plt    # For data visualizations
from sklearn import linear_model   # For ML model training

# For dataset scaling
from sklearn.preprocessing import StandardScaler

# Split your dataset into training and testing subsets
from sklearn.model_selection import train_test_split

# To apply Linear Regression model
from sklearn.linear_model import LinearRegression

# To evaluate performance of ML (regression) model
from sklearn.metrics import mean_absolute_error, mean_squared_error
from sklearn.metrics import r2_score

# For saving the trained ML model
import joblib
```

8.3.2.2 Step 2(a) Data Preparation: Loading/Reading the Dataset

Listing 8.16	Step 2(a): Loading and reading the dataset

```
print("=======================================")
print("Step 2: Data Preparation")
print("Step 2 (a): Loading/Reading Dataset")
print("=======================================")

crop_yield_df = pd.read_csv("crop_yield_dataset.csv")
```

8.3.2.3 Step 2(b) – Exploring the dataset using Pandas' and Matplotlib's built-in methods

Listing 8.17	Step 2(b): Dataset exploration using Python Pandas built-in methods

```
print("Crop Yield Dataset Shape (Rows and Columns)")
print("=============================================")
rows, cols = crop_yield_df.shape
print("Dataset Rows:", rows)
print("Dataset Columns:", cols)

print("Datatypes of Records Stored in Each Column")
print("=============================================")
print(crop_yield_df.dtypes)

print("Data Entries in Each Column of Dataset")
print("=============================================")
print(crop_yield_df.count())

print("Concise Summary of Dataset")
print("=============================================")
print(crop_yield_df.info())

print("Dataset Description")
print("=============================================")
print(crop_yield_df.describe())

print("Brief Tabular View of Dataset")
print("=============================================")
print(crop_yield_df.info)
# or simple print(crop_yield_df) can be used

print("Full Tabular View of Dataset")
print("=============================================")
print(crop_yield_df.to_string())
print("Number of Null Values in Each Column")
print("=============================================")
```

```
print(crop_yield_df.isna().sum())

print("Distribution of Dataset Values in Plotted Graphs")
print("==================================================")
```

Output of Listing 8.17:

The self-explanatory output of Listing 8.17 is shown below.

```
========================================
Step 2: Data Preparation
Step 2 (b): Dataset Exploration
========================================
Crop Yield Dataset Shape (Rows and Columns)
================================================
Dataset Rows: 30
Dataset Columns: 7
Datatypes of Records Stored in Each Column
================================================
Crop_id         object
Rainfall        float64
Temperature     float64
Fertilizer      float64
Pesticide       float64
Crop_Yield      float64
Yield_status    object
dtype: object
Data Entries in Each Column of Dataset
================================================
Crop_id         30
Rainfall        26
Temperature     27
Fertilizer      28
Pesticide       28
Crop_Yield      30
Yield_status    30
dtype: int64
Concise Summary of Dataset
================================================
<class 'pandas.core.frame.DataFrame'>
RangeIndex: 30 entries, 0 to 29
Data columns (total 7 columns):
 #    Column        Non-Null Count    Dtype
---   ---------     --------------    -------
 0    Crop_id       30 non-nul        object
 1    Rainfall      26 non-null       float64
 2    Temperature   27 non-null       float64
 3    Fertilizer    28 non-null       float64
 4    Pesticide     28 non-null       float64
 5    Crop_Yield    30 non-null       float64
 6    Yield_status  30 non-null       object
dtypes: float64(5), object(2)
memory usage: 1.8+ KB
None
Dataset Description
================================================
```

```
          Rainfall      Temperature   Fertilizer   Pesticide   Crop_Yield
count     26.000000     27.00000      28.000000    28.000000   30.000000
mean      834.994615    22.78037      128.345714   3.503571    16.501333
std       178.857913    4.02676       22.751320    1.337832    0.899896
min       608.670000    10.00000      87.750000    2.390000    15.140000
25%       747.432500    22.10000      110.492500   2.877500    15.952500
50%       781.795000    23.62000      130.760000   3.215000    16.490000
75%       860.995000    24.43500      139.580000   3.437500    16.760000
max       1500.000000   28.38000      180.000000   8.030000    19.440000
Brief Tabular View of Dataset
==========================================
<bound method DataFrame.info of Crop_id Rainfall Temperature ... Pesticide Crop_Yield
Yield_status
0      Crop1   849.67    21.17 ...  2.59    15.14    Good
1      Crop1   786.17    23.16 ...  2.72    16.06    Better
2      Crop1   864.77    23.31 ...  3.37    17.34    Optimal
3      Crop1   952.30    22.40 ...  3.31    17.48    Optimal
4      Crop1   NaN       23.68 ...  2.99    15.60    Good
5      Crop1   776.59    24.81 ...  3.06    16.69    Better
6      Crop1   957.92    27.77 ...  3.64    16.85    Better
7      Crop1   876.74    24.35 ...  2.70    17.07    Optimal
8      Crop1   753.05    24.52 ...  3.27    16.92    Better
9      Crop1   NaN       23.85 ...  2.90    18.57    Optimal
10     Crop2   753.66    20.16 ...  2.89    15.75    G
11     Crop2   753.43    23.95 ...  3.55    16.52    B
12     Crop2   824.20    10.00 ...  5.43    16.58    B
13     Crop2   608.67    NaN ...    8.03    15.87    G
14     Crop2   1500.00   23.62 ...  3.65    15.99    G
15     Crop2   743.77    11.00 ...  3.01    16.43    B
16     Crop2   NaN       23.93 ...  NaN     16.71    B
17     Crop2   831.42    21.66 ...  2.84    15.94    G
18     Crop2   709.20    NaN ...    3.16    16.54    B
19     Crop2   658.77    25.50 ...  7.56    16.08    B
20     Crop3   946.56    25.58 ...  3.05    19.44    Optimal
21     Crop3   777.42    22.18 ...  3.30    16.22    Better
22     Crop3   806.75    26.81 ...  2.59    16.34    Better
23     Crop3   1200.00   21.20 ...  NaN     15.41    Good
24     Crop3   745.56    NaN ...    2.50    16.46    Better
25     Crop3   NaN       28.38 ...  2.39    16.76    Better
26     Crop3   684.90    22.02 ...  3.58    15.26    Good
27     Crop3   837.57    22.87 ...  3.40    16.56    Better
28     Crop3   739.94    24.20 ...  3.31    16.76    Better
29     Crop3   770.83    22.99 ...  3.31    15.70    Good
[30 rows x 7 columns]>
Full Tabular View of Dataset
==========================================
Crop_id  Rainfall  Temperature    Fertilizer  Pesticide  Crop_Yield
Yield_status
0        Crop1     849.67   21.17     127.16   2.59      15.14     Good
1        Crop1     786.17   23.16     131.22   2.72      16.06     Better
2        Crop1     864.77   23.31     141.66   3.37      17.34     Optimal
3        Crop1     952.30   22.40     141.08   3.31      17.48     Optimal
4        Crop1     NaN      23.68     92.45    2.99      15.60     Good
5        Crop1     776.59   24.81     101.24   3.06      16.69     Better
```

```
6    Crop1  957.92   27.77  130.30  3.64  16.85  Better
7    Crop1  876.74   24.35  130.28  2.70  17.07  Optimal
8    Crop1  753.05   24.52  130.30  3.27  16.92  Better
9    Crop1  NaN      23.85  NaN     2.90  18.57  Optimal
10   Crop2  753.66   20.16  131.42  2.89  15.75  G
11   Crop2  753.43   23.95  142.71  3.55  16.52  B
12   Crop2  824.20   10.00  139.08  5.43  16.58  B
13   Crop2  608.67   NaN    133.03  8.03  15.87  G
14   Crop2  1500.00  23.62  177.00  3.65  15.99  G
15   Crop2  743.77   11.00  135.18  3.01  16.43  B
16   Crop2  NaN      23.93  104.54  NaN   16.71  B
17   Crop2  831.42   21.66  115.26  2.84  15.94  G
18   Crop2  709.20   NaN    110.29  3.16  16.54  B
19   Crop2  658.77   25.50  121.64  7.56  16.08  B
20   Crop3  946.56   25.58  166.29  3.05  19.44  Optimal
21   Crop3  777.42   22.18  NaN     3.30  16.22  Better
22   Crop3  806.75   26.81  133.73  2.59  16.34  Better
23   Crop3  1200.00  21.20  87.75   NaN   15.41  Good
24   Crop3  745.56   NaN    110.56  2.50  16.46  Better
25   Crop3  NaN      28.38  141.78  2.39  16.76  Better
26   Crop3  684.90   22.02  180.00  3.58  15.26  Good
27   Crop3  837.57   22.87  98.45   3.40  16.56  Better
28   Crop3  739.94   24.20  105.69  3.31  16.76  Better
29   Crop3  770.83   22.99  133.59  3.31  15.70  Good
Number of Null Values in Each Column
================================================
Crop_id 0
Rainfall 4
Temperature 3
Fertilizer 2
Pesticide 2
Crop_Yield 0
Yield_status 0
dtype: int64
```

Listing 8.18	Step 2(b): Dataset exploration using Python Matplotlib methods

```
# Plotting graphs in 2 rows and 3 colums
fig, axis = plt.subplots(2, 3, figsize = (13, 13))

axis[0,0].hist(crop_yield_df['Rainfall'], bins=13, color='green',
edgecolor='black')
axis[0,0].set_title('Distribution of Rainfall Values (in mm)')
axis[0,0].set_xlabel('Groups')
axis[0,0].set_ylabel('Count')

axis[0,1].hist(crop_yield_df['Temperature'], bins=13,
color='orange', edgecolor='black')
axis[0,1].set_title('Distribution of Temperature Values (in
Celsius)')
axis[0,1].set_xlabel('Groups')
axis[0,1].set_ylabel('Count')
```

```
axis[0,2].hist(crop_yield_df['Fertilizer'], bins=13,
color='blue', edgecolor='black')
axis[0,2].set_title('Distribution of Fertilizer Values (in
Kg/ha)')
axis[0,2].set_xlabel('Groups')
axis[0,2].set_ylabel('Count')

axis[1,0].hist(crop_yield_df['Pesticide'], bins=13,
color='skyblue', edgecolor='black')
axis[1,0].set_title('Distribution of Pesticide Values (in
Kg/ha)')
axis[1,0].set_xlabel('Groups')
axis[1,0].set_ylabel('Count')

axis[1,1].hist(crop_yield_df['Yield_status'], bins=13,
color='coral', edgecolor='black')
axis[1,1].set_title('Distribution of Yield Status Values')
axis[1,1].set_xlabel('Groups')
axis[1,1].set_ylabel('Count')

axis[1,2].hist(crop_yield_df['Crop_Yield'], bins=13,
color='plum', edgecolor='black')
axis[1,2].set_title('Distribution of Crop Yield Values (in
Tons/ha)')
axis[1,2].set_xlabel('Groups')
axis[1,2].set_ylabel('Count')

plt.show()
```

Output of Listing 8.18

The output of Listing 8.18 is shown in Fig. 8.11.

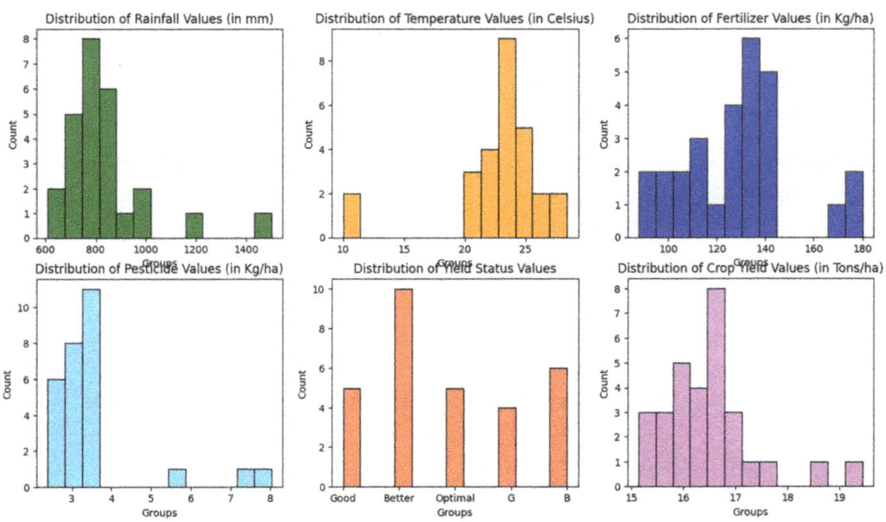

Fig. 8.11 Distribution of values in given dataset columns

8.3.2.4 Step 2(c) Data Preparation: Dataset Cleaning

8.3.2.4.1 Dataset Cleaning by Handling Missing Values

During data exploration in Step 2(b), you may notice that some values in different columns are missing. This can happen for various reasons, such as errors during data collection or entry. Deletion and imputation are two common approaches to handle missing values in a dataset, each having its own advantages and disadvantages depending on the extent and importance of the missing data. In the deletion method, data can be removed either row-wise or column-wise. In contrast, imputation involves filling in the missing entries with estimated values derived from the existing data in the same column. These estimates can be based on statistical measures such as the mean, median, or mode, or more advanced techniques such as regression models. Listing 8.19 demonstrates how missing values can be replaced using the mean of the corresponding column.

Listing 8.19	Step 2(c): Dataset cleaning by handling missing values

```
print("=====================================")

print("Step 2: Data Cleaning")
print("Step 2 (c): Handling Missing Values")
print("=====================================")

# Finding column names in dataset with Missing Values
print("Finding Column(s) with Missing (NaN) Values")
print("=========================================")
print(crop_yield_df.isna().sum())

'''
After identifying the missing values using the isna() method and
knowing which columns have these missing entries, use the fillna()
method to replace the NaN values in the dataset.
'''

# Variable to store column's mean value
col_mean_val = 0.0

# Fill missing values in Rainfall column with its mean
col_mean_val = round(crop_yield_df['Rainfall'].mean(), 2)
crop_yield_df['Rainfall'] = crop_yield_df['Rainfall'].
fillna(col_mean_val)
```

```
# Fill missing values in Temperature column with its mean
col_mean_val = round(crop_yield_df['Temperature'].mean(), 2)
crop_yield_df['Temperature'] = crop_yield_df['Temperature'].
fillna(col_mean_val)

# Fill missing values in Fertilizer column with its mean
col_mean_val = round(crop_yield_df['Fertilizer'].mean(), 2)
crop_yield_df['Fertilizer'] = crop_yield_df['Fertilizer'].
fillna(col_mean_val)

# Fill missing values in Rainfall column with its mean
col_mean_val = round(crop_yield_df['Pesticide'].mean(), 2)
crop_yield_df['Pesticide'] = crop_yield_df['Pesticide'].
fillna(col_mean_val)

# Rechecking number of NaN values in the dataset
print("Rechecking Number of Missing (NaN) Values in Each Column")
print("=============================================================")
print(crop_yield_df.isna().sum())
```

Output of Listing 8.19:

The self-explanatory output of Listing 8.19 is shown below.

```
Step 2: Data Cleaning
Step 2 (c): Handling Missing Values
========================================
Finding Column(s) with Missing (NaN) Values
============================================
Crop_id 0
Rainfall 4
Temperature 3
Fertilizer 2
Pesticide 2
Crop_Yield 0
Yield_status 0
dtype: int64
Rechecking Number of Missing (NaN) Values in Each Column
=========================================================
Crop_id 0
Rainfall 0
Temperature 0
Fertilizer 0
Pesticide 0
Crop_Yield 0
Yield_status 0
dtype: int64
```

8.3.2.4.2 Dataset Cleaning by Detecting and Treating Outliers

Outliers are values that significantly differ from others within the dataset and can be identified during the data exploration process, particularly in Step 2(b). Similar to how missing values are handled, outliers can be addressed by either deleting the affected rows or columns or by replacing them through imputation. Once appropriate upper and lower threshold limits for the dataset column have been defined, this deletion and imputation can be done using statistical measures such as the mean, median, or mode, or regression models as shown in Listing 8.20.

Listing 8.20	Step 2(c): Dataset cleaning by treating outliers

```
print("==============================================")
print("Step 2: Data Cleaning")
print("Step 2 (c): Detection and Treating Outliers")
print("==============================================")

# Checking Outlier possibility using threshold value(s)
print("Detecting outliers in different columns of the dataset")
print("========================================================")
print("Detection and Replacement of Outliers have been Started")
print("...")

# Detect outliers for Rainfall column
outliers = (crop_yield_df['Rainfall'] < 600) | (crop_yield_
df['Rainfall'] >= 1000)
# Compute mean of valid (non-outlier) values
col_mean_val = crop_yield_df.loc[~outliers, 'Rainfall'].mean()
# Replace outliers with the mean
crop_yield_df.loc[outliers, 'Rainfall'] = col_mean_val

# Detect outliers for Temperature column
outliers = (crop_yield_df['Temperature'] < 600) | (crop_yield_
df['Temperature'] >= 1000)
# Compute mean of valid (non-outlier) values
col_mean_val = crop_yield_df.loc[~outliers, 'Temperature'].mean()
# Replace outliers with the mean
crop_yield_df.loc[outliers, 'Temperature'] = col_mean_val

# Detect outliers for Fertilizer column
outliers = (crop_yield_df['Fertilizer'] < 600) | (crop_yield_
df['Fertilizer'] >= 1000)
# Compute mean of valid (non-outlier) values
col_mean_val = crop_yield_df.loc[~outliers, 'Fertilizer'].mean()
# Replace outliers with the mean
crop_yield_df.loc[outliers, 'Fertilizer'] = col_mean_val
```

```
# Detect outliers for Pesticide column
outliers = (crop_yield_df['Pesticide'] < 600) | (crop_yield_
df['Pesticide'] >= 1000)
# Compute mean of valid (non-outlier) values
col_mean_val = crop_yield_df.loc[~outliers, 'Pesticide'].mean()
# Replace outliers with the mean
crop_yield_df.loc[outliers, 'Pesticide'] = col_mean_val

print("Detection/Replacement of Outliers is Completed.")
print("==============================================")

print("Verifying No Outliers in Each Dataset Column")
print("==============================================")

# Checking outliers for Rainfall column
if ((crop_yield_df['Rainfall'] < 600) | (crop_yield_
df['Rainfall'] >= 1000)).any():
    print("Yes Outlier Values in Rainfall Column")
else:
    print("No Outlier Values in Rainfall Column")

# Checking outliers for Temperature column
if ((crop_yield_df['Temperature'] < 20) | (crop_yield_
df['Temperature'] >= 30)).any():
    print("Yes Outlier Values in Temperature Column")
else:
    print("No Outlier Values in Temperature Column")

# Checking outliers for Fertilizer column
if ((crop_yield_df['Fertilizer'] < 85) | (crop_yield_
df['Fertilizer'] >= 150)).any():
    print("Yes Outlier Values in Fertilizer Column")
else:
    print("No Outlier Values in Fertilizer Column")

# Checking outliers for Pesticide column
if ((crop_yield_df['Pesticide'] < 2) | (crop_yield_
df['Pesticide'] >= 5)).any():
    print("Yes Outlier Values in Pesticide Column")
else:
    print("No Outlier Values in Pesticide Column")
```

Output of Listing 8.20:

The self-explanatory output of Listing 8.20 is shown below.

```
==========================================
Step 2: Data Cleaning
Step 2 (c): Detection and Treating Outliers
==========================================
Detecting outliers in different columns of the dataset
==========================================
Detection and Replacement of Outliers have been Started
...
Detection and Replacement of Outliers have been Completed.
==========================================
Verifying No Outliers in Each Dataset Column
==========================================
No Outlier Values in Rainfall Column
No Outlier Values in Temperature Column
No Outlier Values in Fertilizer Column
No Outlier Values in Pesticide Column
```

8.3.2.4.3 Dataset Cleaning by Resolving Inconsistencies

During the data exploration phase, both the tabular view of the DataFrame and the graph showing the distribution of yield status values reveal inconsistencies in the way these values are recorded. For instance, while crops 1 and 3 use the full labels "Good," "Better," and "Optimal," crop 2 uses abbreviated forms "G," "B," and "O" to represent the same statuses, respectively. To ensure consistency across the dataset, these abbreviated entries should be replaced with their full forms, i.e., "G" with "Good," "B" with "Better," and "O" with "Optimal," as illustrated in Listing 8.21.

Listing 8.21	Step 2 (c): Dataset cleaning by resolving any inconsistencies

```
print("=======================================")
print("Step 2: Data Cleaning")
print("Step 2 (c): Detection and Handling Inconsistencies")
print("=======================================")

print("Original Yield Status Values for Crop 2")
print("=======================================")
print(crop_yield_df[crop_yield_df['Crop_id'] == "Crop2"]
['Yield_status'])
```

```
# Handle data inconsistencies detected in Data Exploration
crop_yield_df['Yield_status'] = crop_yield_df['Yield_status'].
replace("G", "Good")
crop_yield_df['Yield_status'] = crop_yield_df['Yield_status'].
replace("B", "Better")
crop_yield_df['Yield_status'] = crop_yield_df['Yield_status'].
replace("O", "Optimal")

print("Modified Yield Status Values for Crop 2")
print("========================================")
print(crop_yield_df[crop_yield_df['Crop_id'] == "Crop2"]
['Yield_status'])
```

Output of Listing 8.21:

The self-explanatory output of Listing 8.21 is shown below.

```
=====================================================
Step 2: Data Cleaning
Step 2 (c): Detection and Handling Inconsistencies
=====================================================
Original Yield Status Values for Crop 2
========================================
10      G
11      B
12      B
13      G
14      G
15      B
16      B
17      G
18      B
19      B
Name: Yield_status, dtype: object
Modified Yield Status Values for Crop 2
========================================
10      Good
11      Better
12      Better
13      Good
14      Good
15      Better
16      Better
17      Good
18      Better
19      Better
Name: Yield_status, dtype: object
```

8.3.2.5 Step 2(c) Data Preparation: Preprocessing

8.3.2.5.1 Scaling Numerical Values

Although in this particular case study there is no need to scale numerical features, it is good to understand what scaling is, why it is needed, and how it can be applied using the *StandardScaler* module of the Scikit-learn library. Scaling is used to scale numerical features of a dataset, and it is very helpful for machine learning models that are sensitive to features' scale, i.e., Logistic Regression, SVM, KNN, and so on. Scaling is required because sometimes in the dataset the columns have different scales. For example, the dataset shown in Table 8.1 has columns on very different scales, i.e., rainfall and fertilizer in hundreds, temperature in twenties, and pesticide around 2–3 and in such cases, if you do not scale, models may give more importance to features with larger values (such as rainfall and fertilizer), even if they are not more important than other parameters. To scale the dataset, you can use the StandardScalar method of the Scikit-learn's preprocessing module, as demonstrated in Listing 8.22.

Listing 8.22	Step 2(d): Dataset scaling

```
print("========================================")
print("Step 2: Data Cleaning")
print("Step 2 (d): Dataset Scaling")
print("========================================")

column_to_scale = ['Rainfall']

# Initialize StandardScaler
scaler = StandardScaler()

# Fit and transform only selected columns
scaled_data = scaler.fit_transform(crop_yield_df[column_to_scale])

# Convert scaled data back to a DataFrame
scaled_df = pd.DataFrame(scaled_data, columns=column_to_scale)
print(scaled_df)
```

Output of Listing 8.22:

The self-explanatory output of Listing 8.22 is shown below.

```
================================================
Step 2: Data Cleaning
Step 2 (d): Dataset Scaling
================================================
Sample Scaled Rainfall Column of Crop Yield Dataset
================================================
         Rainfall
0        0.652866
1        -0.152716
2        0.844430
3        1.954864
4        0.466631
```

8.3.2.6 Step 2(d) Data Preparation: Splitting Dataset

Listing 8.23	Step 2(e): Splitting the dataset into training and testing subsets

```
print("==============================================")
print("Step 2: Data Cleaning")
print("Step2(e): Splitting training & testing subsets")
print("==============================================")

# Features (X) and target (y)
X = crop_yield_df[['Rainfall', 'Temperature', 'Fertilizer']]
y = crop_yield_df['Crop_Yield']

'''
Split the dataset into training and testing subsets (80%
training, 20% testing)
X: features dataset – the input variables (e.g., Rainfall,
Temperature, Fertilizer).
y: target dataset – the output variable you want to predict
(e.g., Crop Yield)

test_size=0.2: Means 20% of the data will be used for testing,
and the remaining 80% for training. You can adjust this value
(e.g., 0.3 for 30% test data, and the remaining 70% for
training).

random_state=42:
This is a seed for random number generation. It ensures you get
the same split every time you run the code.If you want to change
the split then you can use any number other than 42.
'''

X_train, X_test, y_train, y_test = train_test_split(X, y, test_
size = 0.2, random_state = 42)

# Display Split dataset details
print("Training Features:\n", X_train)
print("Testing Features:\n", X_test)
print("Training Target:\n", y_train)
print("Testing Target:\n", y_test)
```

Output of Listing 8.23:

The self-explanatory output of Listing 8.23 is shown below.

```
=====================================================
Step 2: Data Cleaning
Step 2 (e): Splitting the dataset into training and testing subsets
=====================================================
Training Features:
        Rainfall        Temperature     Fertilizer
28      739.940000      24.2000         105.690000
24      745.560000      22.7800         110.560000
12      824.200000      23.6575         139.080000
0       849.670000      21.1700         127.160000
4       834.990000      23.6800         92.450000
16      834.990000      23.9300         104.540000
5       776.590000      24.8100         101.240000
13      608.670000      22.7800         133.030000
11      753.430000      23.9500         142.710000
22      806.750000      26.8100         133.730000
1       786.170000      23.1600         131.220000
2       864.770000      23.3100         141.660000
25      834.990000      28.3800         141.780000
3       952.300000      22.4000         141.080000
21      777.420000      22.1800         128.350000
26      684.900000      22.0200         123.225556
18      709.200000      22.7800         110.290000
29      770.830000      22.9900         133.590000
20      946.560000      25.5800         123.225556
7       876.740000      24.3500         130.280000
10      753.660000      20.1600         131.420000
14      798.207857      23.6200         123.225556
19      658.770000      25.5000         121.640000
6       957.920000      27.7700         130.300000
Testing Features:
        Rainfall        Temperature     Fertilizer
27      837.570000      22.8700         98.45
15      743.770000      23.6575         135.18
23      798.207857      21.2000         87.75
17      831.420000      21.6600         115.26
8       753.050000      24.5200         130.30
9       834.990000      23.8500         128.35
Training   Target:
 28         16.76
24          16.46
12          16.58
0           15.14
4           15.60
16          16.71
5           16.69
13          15.87
11          16.52
22          16.34
1           16.06
2           17.34
```

```
25        16.76
3         17.48
21        16.22
26        15.26
18        16.54
29        15.70
20        19.44
7         17.07
10        15.75
14        15.99
19        16.08
6         16.85
Name: Crop_Yield, dtype: float64
Testing Target:
 27       16.56
15        16.43
23        15.41
17        15.94
8         16.92
9         18.57
Name: Crop_Yield, dtype: float64
```

8.3.2.7 Step 3: Model Selection, Training, and Predictions

Choose an appropriate machine learning algorithm based on the type of problem, i.e., regression analysis, classification, or clustering. After model selection, use the model to train the selected model using the *fit()* method on the training dataset. Ultimately, the trained model can be used to predict outcomes on the test dataset.

For this case study, which is related to predicting crop yield based on one or more input variables (e.g., rainfall, temperature, fertilizer usage), the ideal choice is the Linear Regression Machine Learning Model. Linear Regression works well for both single-variable regression (one feature like rainfall) and multiple-variable regression (rainfall, temperature, and fertilizer together) as shown in Listings 8.24 and 8.25.

Listing 8.24	Step 3: Model selection, training, and predictions

```
# Feature to train model
X = crop_yield_df[['Rainfall']]
# Defining of target variable
y = crop_yield_df['Crop_Yield']

# Dataset splitting (training and testing) step
X_train, X_test, y_train, y_test = train_test_split(X, y, test_
size = 0.2, random_state = 42)
# Creation of Linear Regression model
model = LinearRegression()
```

```
'''
Training of Linear Regression model using fit() method
Model training involves feeding it with data so it can learn the
underlying patterns.
fit() method adjusts the parameters of the model based on selected
feature and target vector.
'''
model.fit(X_train, y_train)

# Using trained model, prediction of results on test data
y_pred = model.predict(X_test)

# Predicted results as Numpy array
print(np.round(y_pred, 2))
# Displaying (converted NumPy array of) predicted DataFrame
print(y_test.to_numpy())
```

Output of Listing 8.24:

The self-explanatory output of Listing 8.24 is shown below.

```
==================================================
Step 3: Model Selection, Training, and Prediction
==================================================
Predicted Crop Yield
[16.7 16.16 16.47 16.66 16.21 16.68]
Actual Crop Yield
[16.56 16.43 15.41 15.94 16.92 18.57]
```

8.3.2.8 Step 4: Model Evaluation

Assess the model's performance using suitable metrics such as Mean Squared Error (MSE), Mean Absolute Error (MAE), R^2 Score, Accuracy, Precision, Recall, F1-score, or Confusion Matrix. As the task in this example is related to regression (i.e., predicting numerical (crop yield) values; therefore, in this case, you have to use regression metrics (i.e., MSE, MAE, or R^2 Score) rather than using classification metrics (i.e., Accuracy, Precision, Recall, F1-score, or Confusion Matrix). The evaluation of predicted regression results is shown in Listing 8.25.

Listing 8.25	Step 4: Model evaluation

```
print("==========================================")
print("Step 3: Model Selection, Training, and Prediction")
print("==========================================")

# Feature (Rainfall) to train model
X = crop_yield_df[['Rainfall']]
# Defining of target variable
y = crop_yield_df['Crop_Yield']

# Dataset splitting (training and testing) step
X_train, X_test, y_train, y_test = train_test_split(X, y,
                                    test_size = 0.2, random_state = 42)

# Creation of Linear Regression model
model = LinearRegression()

'''
Training of Linear Regression model using fit() method
Model training involves feeding it with data so it can learn the
underlying patterns.
fit() method adjusts the parameters of the model based on selected
feature and target vector.
'''
model.fit(X_train, y_train)

# Using trained model, prediction of results on test data
y_pred = model.predict(X_test)

print(f"\nResults using Rainfall")
print("=========================")
# Predicted results as Numpy array
print("Predicted Crop Yield")
print(np.round(y_pred, 2))
# Displaying the converted NumPy array from DataFrame
print("Actual Crop Yield")
print(y_test.to_numpy())

# Predicted Results Evaluation
print("MAE:", round(mean_absolute_error(y_test, y_pred), 2))
print("MSE:", round(mean_squared_error(y_test,y_pred), 2))
print("R2 Score:", round(r2_score(y_test, y_pred), 2))
```

```
# Feature (Temperature) to train model
X = crop_yield_df[['Temperature']]
# Defining of target variable
y = crop_yield_df['Crop_Yield']

# Dataset splitting (training and testing) step
X_train, X_test, y_train, y_test = train_test_split(X, y,
                                   test_size = 0.2, random_state = 42)

# Creation of Linear Regression model
model = LinearRegression()
model.fit(X_train, y_train)

# Using trained model, prediction of results on test data
y_pred = model.predict(X_test)

print(f"\nResults using Temperature")
print("=============================")
# Predicted results as Numpy array
print("Predicted Crop Yield")
print(np.round(y_pred, 2))
# Displaying the converted NumPy array from DataFrame
print("Actual Crop Yield")
print(y_test.to_numpy())

# Predicted Results Evaluation
print("MAE:", round(mean_absolute_error(y_test, y_pred), 2))
print("MSE:", round(mean_squared_error(y_test,y_pred), 2))
print("R2 Score:", round(r2_score(y_test, y_pred), 2))

# Feature (Fertilizer) to train model
X = crop_yield_df[['Fertilizer']]
# Defining of target variable
y = crop_yield_df['Crop_Yield']

# Dataset splitting (training and testing) step
X_train, X_test, y_train, y_test = train_test_split(X, y,
                                   test_size = 0.2, random_state = 42)

# Creation of Linear Regression model
model = LinearRegression()
model.fit(X_train, y_train)

# Using trained model, prediction of results on test data
y_pred = model.predict(X_test)
```

```
print(f"\nResults using Fertilizer")
print("===========================")
# Predicted results as Numpy array
print("Predicted Crop Yield")
print(np.round(y_pred, 2))
# Displaying the converted NumPy array from DataFrame
print("Actual Crop Yield")
print(y_test.to_numpy())

# Predicted Results Evaluation
print("MAE:", round(mean_absolute_error(y_test, y_pred), 2))
print("MSE:", round(mean_squared_error(y_test, y_pred), 2))
print("R2 Score:", round(r2_score(y_test, y_pred), 2))

# All features to train model
X = crop_yield_df[['Rainfall','Temperature','Fertilizer',
                                                  'Pesticide']]
# Defining of target variable
y = crop_yield_df['Crop_Yield']

# Dataset splitting (training and testing) step
X_train, X_test, y_train, y_test = train_test_split(X, y, test_
size = 0.2, random_state = 42)

# Creation of Linear Regression model
model = LinearRegression()
model.fit(X_train, y_train)

# Using trained model, prediction of results on test data
y_pred = model.predict(X_test)

print(f"\nResults using All Features Together")
print("====================================")
# Predicted results as Numpy array
print("Predicted Crop Yield")
print(np.round(y_pred, 2))
# Displaying the converted NumPy array from DataFrame
print("Actual Crop Yield")
print(y_test.to_numpy())

# Predicted Results Evaluation
print("MAE:", round(mean_absolute_error(y_test, y_pred), 2))
print("MSE:", round(mean_squared_error(y_test, y_pred), 2))
print("R2 Score:", round(r2_score(y_test, y_pred), 2))
```

Output of Listing 8.25:

The output of Listing 8.25 is shown below.

```
==================================================
Step 3: Model Selection, Training, and Prediction
==================================================
Results using Rainfall
=========================
Predicted Crop Yield
[16.7 16.16 16.47 16.66 16.21 16.68]
Actual Crop Yield
[16.56 16.43 15.41 15.94 16.92 18.57]
MAE: 0.8
MSE: 0.97
R2 Score: 0.01
Results using Temperature
============================
Predicted Crop Yield
[16.29 16.43 15.97 16.06 16.6 16.47]
Actual Crop Yield
[16.56 16.43 15.41 15.94 16.92 18.57]
MAE: 0.56
MSE: 0.82
R2 Score: 0.16
Results using Fertilizer
===========================
Predicted Crop Yield
[16.29 16.53 16.22 16.4 16.5 16.49]
Actual Crop Yield
[16.56 16.43 15.41 15.94 16.92 18.57]
MAE: 0.69
MSE: 0.91
R2 Score: 0.07
Results using All Features Together
=====================================
Predicted Crop Yield
[16.59 16.17 16.21 16.32 16.36 16.62]
Actual Crop Yield
[16.56 16.43 15.41 15.94 16.92 18.57]
MAE: 0.66
MSE: 0.83
R2 Score: 0.15
```

Explanation:
Although the main purpose of this code is to demonstrate how to implement machine learning models, a brief explanation of the results (below) is provided below to provide some basic understanding.
The linear regression results indicate that none of the individual features—rainfall, temperature, or fertilizer can strongly predict crop yield, as shown by the low R^2 scores (0.01–0.16). Among them, temperature performs slightly better ($R^2 = 0.16$, MAE = 0.56), suggesting it has a relatively stronger relationship with yield compared to the other single features. Moreover, it is interesting to note that combining all three features does not significantly improve the model ($R^2 = 0.15$), indicating that either the features are not highly informative or the relationships between features and yield are not well captured by a linear model. To improve performance, you may need to enrich the dataset with more records or use more diverse variables (e.g., soil quality, crop variety).

8.3.2.9 Step 5: Saving the Trained Model

Save the final model for future reuse using the Python joblib module.

Listing 8.26	Step 5: Saving trained machine learning model

```
# Saving trained machine learning model
print("=====================================")
print("Step 5: Saving Machine Learning Model")
print("=====================================")

# Save the model to a file
joblib.dump(model, 'crop_yield_model.joblib')
print("Machine Learning Model Saved Correctly!")

# To load the model later
loaded_model = joblib.load('crop_yield_model.joblib')
```

Output of Listing 8.26:

The result of this execution will be a saved file stored on the specified path. However, the self-explanatory output of Listing 8.26 on the console is shown below.

```
=====================================
Step 5: Saving Machine Learning Model
=====================================
Machine Learning Model Saved Correctly!
```

8.4 Exercises

Problem 8.1 Write a Python program to plot line and bar graphs of the following yield trends of crops (wheat, maize, barley, and oats) at an agricultural farm from 2020 to 2024.

```
wheat_yield = [4.5, 3.7, 3.8, 4.7, 5.2]
maize_yield = [4.0, 2.1, 2.9, 3.9, 3.9]
barley_yield = [3.0, 3.1, 3.3, 3.5, 3.6]
oat_yield = [5.0, 2.1, 3.9, 3.5, 4.6]
```

Problem 8.2 Write a Python program to visualize the relationship between Soil pH and Crop Yield in a scatter plot while considering the following dataset:

```
soil_ph = [5.5, 6.0, 6.5, 7.0, 7.5, 8.0]
crop_yield = [2.5, 3.0, 3.5, 3.8, 3.0, 2.2] # tons/ha
```

Problem 8.3 Considering the dataset shown in Table 8.2, use machine learning techniques to build predictive models that estimate milk yield using:

- Single feature at a time, for example,
  - Predict milk yield based on livestock Weight
  - Predict milk yield based on livestock Age
  - Predict milk yield based on livestock Temperature
- Multiple features collectively, for example,
  - Predict milk yield using a combination of Weight, Age, and Temperature of livestock

Table 8.2 Livestock milk production dataset

Weight	Age	Temperature	Milk_Yield
450	2	38.5	15.2
500	3	39	18
520	4	39.2	14.5
470	2	38.7	16
490	5	40.1	12.3
530	3	38.9	17.1
510	4	39.3	13.7
480	2	38.6	16.5
495	5	39.5	11.9
515	3	38.8	17.8

Index

A
abs(number), 120
add(), 174
Aggregation functions, 215–216
Agriculture, 1–4, 6, 8–9, 27, 34, 38, 46, 131, 205, 206, 208, 219, 225, 226, 242, 255, 256, 275–277
append(), 152
Argument passing, 101, 103–106
Argument passing by reference, 104–106
Argument passing by value, 103–104
Arithmetic operators, 27, 37–38, 147
Assignment statements, 34–37
Associativity, 62

B
Bar graphs, 256, 264–271, 300
Binary files, 183, 184, 186, 191–193, 201–203
Boolean expressions, 61, 64–67, 69, 78, 79
Boolean indexing, 215–216
Break statement, 88–89
Built-in function, 114, 120, 143
Built-in Modules, 54, 129–135
Built-in scope, 113

C
capitalize(), 121
ceil(number), 35, 53, 56–58, 60, 134
clear(), 153, 155, 175
Comments, 27–30, 100, 256, 258
Comparison operators, 61–64, 66, 143, 147–149, 212
Compiler, 10, 11, 16, 17, 49
Compound data types, 34, 87, 103, 104, 118, 137
Compound or augmented assignments, 37
Computer programming, 1, 9–11, 95
Computer programming languages, 10, 29, 183
Computer programs, 9–11, 29, 30, 52–54, 61, 64, 88, 109, 183, 185
concat(), 244, 245, 249
Conditional structures, 53, 66–75, 90
Constants, 27, 31–34, 113, 123, 125
Continue statement, 89–90
Control flow structures, 27, 52–54, 61–93
copy(), 153, 156, 202–203
count(), 121, 156, 170
Customized filtering, 241–242

D
Data acquisition, 4, 23
Data analytics, 2, 4, 6, 24
DataFrames, 225, 227–252, 255–264, 289
Data processing, 4–6, 23
Data science, 11, 18, 23, 205–252, 255–301
Data storage, 2, 4, 6, 24
Data transmission, 4–6, 24
Data types, 31–35, 37, 40, 87, 118, 132, 137, 138, 206
Datetime module, 132–133
Deep Copy, 152
Dictionaries, 35, 55, 87, 103, 104, 113, 137, 140, 147, 177–181, 227, 231, 255, 256, 260
Digital agricultural ecosystems, 2, 4, 5
Digital agriculture, 1–9, 23–24, 40, 99, 256
discard(), 175
divmod(), 119, 120

M. A. Iqbal, *Python for Agriculturists*,
https://doi.org/10.1007/978-3-032-01442-9

E
Enclosing (Nonlocal) Scope, 112–113
Enclosing scope, 32, 109, 112–113
endswith(), 121
Error handling, 47–52
Escape sequences, 40–42, 185
eval(), 40, 113, 114
Exception handling, 27, 32, 49–52, 60, 186, 187, 200
Exceptions, 32, 47, 49, 50, 187, 195
Executable statements, 27–31
exp(), 134
Expressions, 27, 34, 35, 37–40, 46, 53, 56, 64, 74, 140
extend(), 152

F
fabs(), 134
File closing, 183
File deleting, 183
File existence check, 183–185
File handling, 24, 32, 55, 113, 129, 183–203
File opening and closing, 183, 185–187
File reading, 183, 187–191
File writing, 183, 193–194
finally, 32, 187
float(), 40, 49, 50, 113, 114
floor(number), 38, 134
for loop, 53, 78, 83–90, 143, 145, 151, 162, 163, 170, 173, 190
format(), 116
Formatting in *print*() function, 116–118
F-strings, 116

G
genfromtxt(), 220–222
Global scope, 110–112

H
Histogram graph, 275–276
Horizontal bar graphs, 265–267

I
Identifiers, 31, 97, 138, 177
if statement, 53, 66, 68, 69, 77
if-elif-else statement, 53
if-else statement, 53, 67, 74
Immutable, 103, 104, 168
Indentation, 12, 27, 30, 47, 66, 76–78
Indentation errors, 76, 77
index(), 156, 170
Indexing and slicing, 40, 42–45, 213–215
Input statements, 39
input() function, 114
insert(), 152
int(), 40, 114
Integrated Development and Learning Environment (IDLE), 17–19, 25
Integrated Development Environment (IDE), 13, 16–18, 21
Interpreter, 10–12, 16–18, 20, 22, 47, 49, 99–102
isalpha(), 121
Iteration Control Flow, 53

J
join(), 121, 244, 245, 248
Joining, 225, 244–249
Jump flow control, 54, 61, 87–91

K
keyword.kwlist, 31
Keyword-based argument passing, 98, 101

L
Left-associative, 62
len(), 113, 121, 146–147, 170
Line graphs, 256–264
List comprehensions, 138, 140–142, 158, 161
Lists, 35, 55, 87, 103, 138, 147, 152, 159, 161, 165
loadtxt(), 220–222
Local Scope, 109–110
Logical errors, 47, 49, 76, 77
Logical operators, 61–64, 66, 67, 78
Loop Control Flow, 53
lower(), 121

M
__main__, 127, 128
Math functions, 120
Math module, 133–135
Mathematical functions, 119–120, 133
Matplotlib, 18, 24, 55, 255–277, 280–285
max(number), 120
Membership testing, 137, 138, 143, 146, 148, 163–165, 170, 173, 179–180
merge(), 244, 245, 248

Merging, 225, 226, 244–249
Methods, 31, 32, 54, 91, 95–136, 152, 153, 155–158, 165, 170–172, 174, 175, 184, 187, 193, 194, 202, 203, 205, 208, 220, 226, 228
min(number), 120
Modules, 24, 27, 31, 32, 54, 92, 95–136, 184, 195, 196, 202, 205, 255, 277, 278, 291, 300
Multiple graphs, 271–272
Mutable, 31, 103–105, 138, 177

N
__name__, 127, 128
Nested if-else statement, 65, 68, 72–74
Non-executable statements, 27–29
Nonlocal, 32, 112, 113
Non-primitive data types, 35
NumPy, 55, 206–225, 255, 276
NumPy array broadcasting, 216–218
NumPy arrays, 206–213, 215, 217–220, 227, 231, 255

O
Operator precedence, 38, 62, 63
Os.rmdir() method, 195–196

P
Pandas, 18, 23, 24, 55, 206, 225–251, 255, 261–262, 277, 280–285
pass statement, 90–91
Pie graphs, 256, 272–274
pop(), 155, 175
Position-based argument passing, 98
pow(), 119, 120
Primitive data types, 34, 37, 55
print() function, 40, 113–118
Programming errors, 47–52
PyCharm, 17, 18, 21–23, 25
Python data structures, 34, 35, 55, 113, 137–181, 256–264
Python functions, 54, 95–136
Python keywords, 31, 32, 47, 97, 113
Python libraries, 55, 255–301
Python method, 96
Python modules, 55, 123–129, 131
Python packages, 14, 23, 24, 55, 205–252, 255, 279
Python programming language, 11–13, 31, 90
Python statements, 15, 16, 27–30, 54, 65

R
randint() method, 130
Random module, 129–132
random() method, 130
Ranking, 223, 242–244
read(), 187, 188, 191, 196, 198, 201
read(integer_chunk_size), 10, 30, 31, 55, 184–193, 196, 198
remove(), 155, 175, 194, 195
Repetition, 46, 53, 61, 78, 170
replace(), 121
Reshaping, 218–220
Resizing, 218–220
return statement, 32, 61, 87, 91, 97, 106–109
reverse(), 153, 156
rmtree() method, 196
round(), 119, 120
Runtime errors, 47–50, 186

S
Scatter graphs, 256, 274–275
Scikit-learn, 24, 55, 255, 276–300
seed() method, 131–132
Selection Flow Control, 53
send2trash() method, 195
Separator parameter, 40, 45
Sequential Flow Control, 53
Series, 53, 225–227
Sets, 35, 55, 87, 103, 261
Shallow Copy, 150, 151
sort(), 153, 156
Sorting, 55, 137, 242–244
split(), 121, 277
sqrt(number), 134
Stacked bar graph, 267–269
startswith(), 121
strftime(), 132
String functions, 120–123
Strings, 35, 40–42, 46, 47, 54, 97, 100, 103, 113, 114, 116, 118, 120, 121, 132, 138, 140, 185, 189, 193
strip(), 121, 189, 190
sum(), 119, 120, 146–147, 170
Syntax errors, 17, 47, 49

T
Ternary statements, 53, 65
Text files, 55, 183, 184, 186–191, 193, 197, 198, 225
Traversing, 143, 145–146, 162–163, 178
try-except-else-finally block, 187

Tuples, 35, 55, 87, 103, 113, 137, 140, 168–172, 181, 255, 256, 259–260
Twin line graph, 263–264
Two-dimensional list, 158–167
Type conversion functions, 118–119

U
update(), 174
upper(), 121
User-defined function, 97–103, 112

V
ValueError, 50–52
Variables, 9, 17, 18, 27, 31–35, 39, 47, 51, 52, 61, 79, 80, 82, 87, 103, 106, 108–114, 127, 128, 170, 172, 185, 294, 299
Variable scope, 109–113
Vertical bar graph, 264–265
Visual Studio Code (VS Code), 17–21, 25

W
while loop, 53, 78–83, 86, 88, 89, 92, 189

Z
ZeroDivisionError, 50, 52

GPSR Compliance
The European Union's (EU) General Product Safety Regulation (GPSR) is a set of rules that requires consumer products to be safe and our obligations to ensure this.

If you have any concerns about our products, you can contact us on

ProductSafety@springernature.com

In case Publisher is established outside the EU, the EU authorized representative is:

Springer Nature Customer Service Center GmbH
Europaplatz 3
69115 Heidelberg, Germany

www.ingramcontent.com/pod-product-compliance
Lightning Source LLC
Chambersburg PA
CBHW051423031125
34911CB00001B/17
9783032014412